Elementary Principles of

Probability
and Information

A

Elementary Principles of

Probability

&

Information

R. F. Wrighton

Academic Press
London & New York
A Subsidiary of Harcourt Brace Jovanovich, Publishers

Elementary Principles of

Probability
and Information

R F WRIGHTON
Medical School,
University of Birmingham,
England

Academic Press
London · New York
A Subsidiary of Harcourt Brace Jovanovich, Publishers

ACADEMIC PRESS INC. (LONDON) LTD.
24/28 Oval Road
London NW1

United States Edition published by
ACADEMIC PRESS INC.
111 Fifth Avenue,
New York, New York 10003

Library of Congress Catalog Card Number: 73–19026
ISBN: 0–12–765550–6

PRINTED IN GREAT BRITAIN BY
ROYSTAN PRINTERS LIMITED
Spencer Court, 7 Chalcot Road
London NW1

Preface

The purpose of the present book is to supply the beginnings of a unified approach, based on a single philosophical principle, to the related fields of probability theory and information theory. I have had in mind throughout the special relevance of each to the biological sciences. Nearly half of the book is directly concerned with inferential problems underlying the use of statistics. The conclusions reached under this heading, however, lend weight to the thesis, widely maintained by other writers on more diffuse grounds, that the statistical approach, especially where regarded as an exemplar of the experimental method, has been, and is likely to be in the future of very minor value. I have briefly indicated in Chapter 7 a more likely direction for fruitful development, in an area which is clouded by the epistemological inadequacies of physical theory. Biologists are currently witnessing a resurgence of anthropomorphism as information notions percolate into biological enquiry conceived at the molecular level, and there seems to be a need for more rigorous ways of thought.

Again regarding future possibilities, I should draw attention to the inadequacy, for wider purposes, of the short discussion in Chapter 6 of the notion of a biological character. In traditional taxonomical enquiry and in clinical medicine, to take two examples, inferences and decisions are frequently made, in part at any rate, on the basis of complex webs of analogy. Doubtless such procedures will always be carried out in the main without the aid of a formal caculus, just as in everyday life we move things about without much recourse to geometrical theorems. Nevertheless, it seems to me not to be beyond the bounds of reasonable possibility to hope that a rational theory of analogy may one day emerge. Although over a long period people have tended to confuse analogy with probability, such a theory would have nothing to do with probability, at least as defined in the following pages. Indeed one could hardly expect it to arise at all within the conventional framework of positivist science.

Any ideas with pretensions to originality have appeared under the author's name in various journals over a period of twenty years or so; the specificaiton of the therapeutic trial in terms of non-replacement sampling in *Acta*

Genetica, the convolution procedure for random sequences in the *British Journal of Social and Preventive Medicine*, the connection between Bayesian theory and Information Theory in a later series of papers in *Acta Genetica*, and the approach to thermodynamical theory outlined in Chapter 7 in the International *Journal of Mathematical Education*. The unified approach to the whole range of problems involved, which is made possible by the discussion at the commencement of Chapter 1 appears here for the first time.

The subject-matter of the book spans a number of topics conventionally regarded as independent, and each possessed of a highly mathematical literature. I do not myself believe that the fundamental issues involved are primarily mathematical, and I have endeavoured to keep mathematical developments to a minimum. I have had, however, to assume some knowledge of the elementary technicalities of probability theory, such as normal approximations, of which an account at an appropriate level is to be found, for example, in T. C. Fry's "Probability Theory and its Engineering Applications" (2nd edition, 1965). For the reader unfamiliar with Information Theory, an introduction to what has come to be regarded as the accepted theory is to be found in H. Quastler's article "A Primer on Information Theory" (*In* "Symposium on Information Theory in Biology", ed. H. P. Yockey, 1958). Similarly, the reader unfamiliar with thermodynamics will find an exceptionally clear and relevant account using very simple mathematics in Gurney's "Introduction to Statistical Mechanics" republished by Dover Books, 1966). Fermi's "Thermodynamics" provides a concise account of classical thermodynamics. I have assumed some degree of acquaintance with the history of probability theory and its relationship to statistical theory. Discussions are readily available in the works of various writers, notably Todhunter, Poincaré, Borel, Keynes, Von Mises, Neyman, Hogben, to mention but a few of the more significant. But a systematic bibliography would be very cumbersome; the recent technical literature is very large, and almost every mathematician and philosopher with broad interests has had something to say on these subjects. The same can be said of more recent discussions of information theory and of its relationship to physics.

The book owes its existence to the good offices of Dr. Louis Opit, to whom I must offer my grateful thanks. I have also received valuable assistance from Dr. Brian Thorpe in checking formulae.

Raymond Wrighton

July, 1973

Contents

1. Classical Random-event Probability

The early eighteenth century Neapolitan philosopher and professor of Rhetoric, Giambattista Vico, was the author of a theory of knowledge which has particular relevance to the two principal themes of the present book. Vico's primary object was to challenge the dominant Cartesianism of his contemporaries*. In place of Descartes' assertion that whatever is clear and distinctly conceived is true, he substitutes *verum ipsum factum*: truth is identical with that which is created. The doctrine to which this apothegm leads, and which it summarises, has the most far-reaching consequences. It destroys the Cartesian hierarchy of the sciences, wherein knowledge becomes less and less certain as it departs from the supposed formalisable self-evidence of geometry. Mathematics retains a special position, since in mathematics Man creates the object of his study, which therefore he wholly understands. Likewise Man may hope to acquire an understanding of comparable depth within the humanities; for he has created his own history, his own language and his own social environment. The Natural Sciences suffer a demotion, relative to the ideal of certainty, and revert to the status which Bacon accorded them; experiment and observation provide, and must always provide the true foundation of physics, since, as Vico puts it, Man can never wholly understand what God has created; physical theory primarily reflects, or epitomises, man's increasing control over his physical environment, rather than an independent reality.

The term *Probability* characteristically relates to mental uncertainty, and so to happenings which we can neither fully predict nor completely control. Most of the things which happen around us, provided we are not misled into selecting special instances, are uncertain in this sense; but since these uncertainties are not of our making, we cannot hope to comprehend them fully, nor partially save by oblique and empirical methods. On the other hand, it is possible for us purposively to create uncertainty. This type of uncertainty we can understand, and in consequence we can develop a true theory to accommodate it. We call this the Theory of Probability. The

* See Croce's "Philosophy of Vico", English translation by R. G. Collingwood, particularly Chapter 1 and Appendix 3.

1

object of the present chapter is to indicate its elementary features by devising a precisely-defined scheme for inducing the occurrence of uncertain events. It turns out that we cannot create ideal uncertainty, but that we can approach it as closely as we choose.

The Nature of Random Events

It is not unusual in the history of Science for what seems to some people to be a childish activity, with no scholarly connections and without economic motivation, to yield an insight which is fundamentally new. There is good reason for this. We may identify play-like activity in any species with the free exploration of capabilities or creativity in a protected environment, and so point to an analogy with the pursuit of pure science. Probability theory began with the mathematical discussion of games of chance, and affords a very special and striking illustration of this theme; for, as we shall see, the creation of uncertainty depends upon the limitations of man's *motor* capabilities which games of chance characteristically exhibit.

When we talk of games of chance, we think of card-shuffling, the tossing of a coin, the throwing of dice, the spinning of a roulette wheel and other such procedures. We need to consider what these have in common, and how they can be rigorously controlled. In the first place we must note that in each case there exists a symmetry in the material structure of the apparatus employed: for example a pack of cards is manufactured so that no card is distinguishable from another by the type of physical procedure employed in shuffling; a die is manufactured in the anticipation that in dynamical respects it will behave like a perfect cube. In the second place, we must note that a controlled latitude is allowed to the operator who manipulates the simple forms of apparatus concerned; thus we require, though it is customary to express the requirement in a very vague fashion, that he must exercise a certain minimum vigour in the shuffling, tossing, spinning or shaking procedures he executes. If the number of facets of symmetry which the apparatus can exhibit is n, we say that, provided such minimum vigour is employed, a *random* event results from the individual *random* trial, and that each of the n possible outcomes takes place with a numerical probability of $1/n$.

If we are to advance beyond the vagueness implicit in the casual origin of the line of enquiry with which we are concerned, we must examine separately the two basic components in procedures for the induction of random events, namely the apparatus employed and the instructions given to the operator. This, however, is just a preliminary step. We shall see that so soon as we depart from the assumption that ideal symmetries can be produced in the artefacts concerned, these two features become inextricably related: any imperfection in the apparatus becomes associated with ambiguity in the instructions to the operator.

Some procedures which are frequently used in games of chance, such as card-shuffling, may be simple and convenient in practice, but from a more theoretical point of view are untidy and unhelpful; we shall discuss first a type of apparatus which requires the very simplest form of instruction to be given to the operator. Let us consider a top made up of a pointed metal rod passing through the centre of a symmetrical hexagonal disc, which we suppose to be perfectly constructed. A random event comprises the resting of the top on one of its labelled edges after spinning by the operator. We can give instructions to the operator which are perfectly precise, in the sense that he palpably violates them if he is able to induce the occurrence of determinate events. All we need do, the top being assumed to be perfect and the table on which it is spun fairly even, is to require that he imparts to it at least an assigned quantity of angular momentum; alternatively, that he spin it so that it certainly executes at least an assigned number of complete rotations. That we thereby provide ourselves with the means for inducing the sort of event we can refer to as "random" depends now on the existence of what we can loosely think of as a *physiological* property of the operator, associated with the limited competence of his motor functioning. Thus we may say that, if we require him to spin the top as indicated we are, in effect, setting him a task, say that of ensuring that the top comes to rest on the edge labelled 6, which he is quite unable to perform at will.

It is sometimes said to be a mere empirical fact that when a coin is tossed it comes down heads with probability one half; and that a suitably-devised machine could toss coins so that they came down heads every time. The suggestion is based on a total misconception. A coin comes down heads with probability one half because we arrange that it should do so: the empirically-verifiable consequences cannot be other than they are. If an operator, within the terms of our instructions to him, were to train himself to toss a coin so that it always came down heads, we should have to regard our instructions as misconceived, and would either have to raise the minimum angular momentum assigned or supply him with a smaller or lighter coin: it is a matter of making the task implicitly assigned to the operator sufficiently difficult. Thus we cannot think of a *random event* without somehow involving a human being in its generation. In this respect we can compare probability theory with arithmetic; the notion of a positive integer depends on the ability which human beings have for singling out elements in their environment which submit consistently to the process of counting.

The Theorem of Bernoulli

At this point, the notion of a random event depends upon the assumption that the apparatus employed in generating it possesses attributes of ideal symmetry. Granted this assumption, we can unambiguously associate with any random event a numerical probability. If an ambiguity is found, it will be traceable to an ambiguity in the specification of the symmetry of the apparatus and will have nothing to do with the behaviour of the operator, that is to say with probabilistic features of the situation. We are led directly to the familiar addition and multiplication laws of formal probability theory. Thus the probability-measure to be attached to the single event comprising the occurrence of either a *five* or *six* in a single throw of an ideal die, is the sum of one sixth and one sixth; the probability that two successive throws of a die will yield first a *five* and then a *six* is the product of one sixth and one sixth. All formal consequences flow, as combinatorial theorems, from these two laws. One such consequence is of especial importance in the present context, since it serves to specify the sense in which ideal probability-measures must be expected to be reflected in observable relative frequencies. This is *Bernoulli's Theorem*, the so-called *weak law of large numbers*. A proof is given in the appendix to the present chapter. The theorem is expressible in purely combinatorial terms; in the language of probability it states, that if a random event occurs with probability p then, given any two numbers ε and δ, it is possible to find a value n (dependent upon ε, δ, and p) such that in n trials the probability that the proportion of occurrences of the event lies between $p - \delta$ and $p + \delta$ is greater than $1 - \varepsilon$. In this form it is comparable with the definition of an ordinary mathematical limit. We say that if n is the number of trials and r the number of successes, r/n converges in probability to p, or that, as the number of trial tends to infinity, the proportion of successes tends, in the special sense defined by the theorem, to the limit p. Bernoulli's theorem underlies any programme devised to examine empirically the assumption that a particular piece of probability apparatus departs to a sufficiently small degree from ideal symmetry to justify the use of the theory, and it underlies any practical use of the theory itself. We note that in any association of probability with observable phenomena, we rely on a convention that some preassigned class of events having very small probability measure can, for practical purposes, be regarded as quite incapable of occurrence.

Imperfections in the Apparatus Employed

We proceed to examine some of the difficulties which arise when we abandon the assumption of perfect symmetry in apparatus used to generate random events. We must first, however, remark on an intrinsic inconsistency which any such assumption carries with it, and which is not as trivial as might first appear. Let us think of a cubical die and suppose, initially, that we can in fact produce one which is perfectly symmetrical in respect of all physical attributes. It might seem to be ideally satisfactory if we could achieve this. We cannot, however, immediately use such a cube to generate random events. To do so we must label the separate faces so that the six essentially different events which can occur as a result of an individual trial are distinguishable. This we can only do by physical means, as for example by painting different numbers of pips on the six faces. In doing this, however, we destroy the physical symmetry. If we are willing to tolerate some departure from ideal symmetry and can regard such inaccuracies as wholly commensurate with the small departures from symmetry induced by sufficiently discreet labelling, we have nothing to worry about. On the other hand, we must anticipate that probability theory as we understand it in the present context will break down if we cannot make this assumption; and indeed this does turn out to be the case at the level of micro-physics. We can conveniently refer to the probability theory which does or can ignore this uncomfortable possibility as the theory of *classical random-event probability*, and it is to this type of probability we shall be confining our attention. Thus, in what immediately follows we shall assume that the asymmetries associated with necessary labelling are quite swallowed up in the imperfections associated with imperfect construction.

We continue to consider dice-throwing. The casual requirement that the operator should shake the die in a pot fairly vigorously is likely to be amply sufficient to ensure that the events induced are random on the assumption of perfect symmetry obtaining. The requirement is equivalent to assigning a minimum value to each component of angular momentum of the die, this having to be exceeded as the operator throws it, or at least at some stage after he has taken his eyes off it. Let us now allow that the die is imperfect. We must expect the imperfections to be reflected in distortions of the relative frequencies of the various possible outcomes in long series of trials, and we can conceive of a programme of experiments designed to explore such deviations systematically. A first suggestion, unthinkingly adopted by many authors, is that a specifiable departure from symmetry has associated with it a unique set of unequal probabilities, each differing from one sixth, but adding up to unity. According to this view, the die possesses these numerical properties in the same way as it possesses a definite volume or a definite specific gravity. It is easy, however, to see that this view is erroneous.

Let us take a wooden die, and heavily load it, say by putting a tack into the face labelled with one pip. The operator shakes the die in a pot, and then throws it onto a table in the way children do. We shall find that in sequences of, say, 1000 throws, the physical bias is reflected with fair consistency in the successive proportions of sixes observed. Suppose, however, that at each trial the operator stands at one end of a large room, and, after agitating the die, throws it high in the air across the room. We shall find that the proportion of sixes rises; the bias, as it were, has more opportunity to exercise its influence. If, on the other hand, the operator agitates the die very vigorously in the pot, then quickly clamps it down on the table in front of him, we shall find that the observable effect of the bias is greatly reduced, and perhaps wholly eliminated. The circumstances in which the die is thrown can thus be expected to influence the relative frequencies observed as soon as we admit a departure from symmetry. Accordingly we must note that, since the nature of the instructions to operator is such that the precise manner of throwing is *not* fully specified, we are implicitly allowing him some control over the circumstances of the individual trial, and therefore over the relative frequencies induced in long sequences of trials. We must suppose, furthermore, that this sort of thing will happen even when the departure is so slight that the number of trials required to demonstrate the effect is prohibitively large. This is a key issue, since for a fundamental theory to be useful it must possess a precision which exceeds that of the means employed in its empirical verification.

We may draw two broad conclusions from this discussion and the essentially experimental findings it presupposes. Firstly, we cannot satisfactorily explain the behaviour of biassed apparatus, that is to say any real apparatus, by merely revising the values of fundamental probabilities in accordance with observed relative frequencies. Secondly, any biassing of the apparatus inevitably affords the operator same opportunity for influencing the long-run, and by implication *any*, outcome of trials.

The second of these conclusions yields an insight into a perennially recurring controversy. People do experiments in which they infer the existence of *extra-sensory* perception from the observation that some subjects can do better than chance would seem to allow in anticipating or matching supposedly random events generated out of sight by an independent operator. We may not be willing to admit that the type of extra-sensory perception envisaged is a meaningful possibility. Let us merely suppose, however, that it does not exist. Then such observations are just what one would expect. Whatever apparatus is employed, it inevitably embodies imperfections. Accordingly, the operator, whether he is aware of it or not, has the opportunity for imposing some sort of underlying pattern on the sequence he generates. On the other hand the subject who is endeavouring to match the operator s results will certainly be generating patterns of his own, since we cannot attribute to him the gift of being able to write down random sequences. Sometimes these patterns will match those of the operator, sometimes they will not; and if they do, the uncritical investigator will credit the subject with extraordinary powers.

Convolution of Random Sequences

We turn our attention now to the main goal of the present chapter. We consider the problem of actually generating sequences which are to a high degree random, whereas hitherto we have only conceived of sequences which are perfectly random in the ideal world of perfectly symmetrical apparatus. Let us review what to this point we can regard as achievable in the way of fabricating truly random sequences. We are limited by two factors. Firstly, we cannot produce perfectly symmetrical apparatus. Secondly, in order to assay the effects of departures from symmetry, we must perform long series of trials; and in doing this we are inhibited to an extent which the law of large numbers indicates. It is difficult to express clearly the precise implications of these limitations. We may say roughly, using the simple example of coin-tossing, that all that experiment can tell us about the sequence induced by an operator tossing a coin according to appropriate instructions is that the coin behaves probabilistically, in an incompletely-definable sense, with probability of heads lying within a range $\frac{1}{2}(1 \pm \delta)$; a realistic value for δ might be something like 1/5*. With this in mind, we can make a more formal statement. It expresses more rigid restrictions than in fact obtain; but, this does not matter, since we shall find that it suffices for the solution of the problem we have to solve. For the sake of convenience, we refer to coin-tossing; but, *mutatis mutandis*, what we say applies equally to any form of apparatus used to generate random events.

We shall assert then, that *in generating a sequence which can be taken to be approximately random, using a nearly symmetrical coin, an operator should be regarded as behaving as if, at each trial, he can choose a probability for the occurrence of heads, as he likes, in a range* $\frac{1}{2}(1 \pm \delta)$.

We shall find it convenient in these circumstances to refer to a "$\frac{1}{2}(1 \pm \delta)$" coin-tossing procedure; and we shall see that the precise value of δ is relatively unimportant. The italicised statement presupposes a scheme of empirical enquiry which establishes both an acceptable value for δ and a value for the preassigned minimum angular momentum which determines the tossing procedure. There exists sufficient such background knowledge for us to regard the conventional tossing of any ordinary coin as falling well within these terms of reference, say with $\delta = 1/5$. When we talk in this context of a sequence, or more accurately of a procedure, which is approximately random, the approximation is very special. Suppose that an operator actually has a perfectly symmetrical coin and tosses it so as to provide a perfectly random sequence, parameter $\frac{1}{2}$; at every hundredth trial, he interposes the result *heads*. We might reasonably say that the resulting sequence is approximately random, particularly if we do not know the phase at which the operator interposes determinate heads. But it would not be the sort of sequence envisaged in the statement given above, and it would be inappropriate to the purposes we have in mind.

* The δ used here is not that employed in discussing Bernoulli's theorem above.

It is necessary to make a preliminary remark about the term *independent* when used of distinct events of a random sequence. If the random events derive from ideally symmetric apparatus, there is no difficulty. Two random events are independent if they proceed from separate acts of volition on the part of the same or different operators. If the apparatus is not perfect, that is to say if we have to admit that the operator can exercise some degree of control over the events he generates, we require a qualification to this definition. We shall say that two successive trials are independent, if the operator is not aware when he performs the second trial of the outcome of the first. In particular, successive tossings of an imperfect coin are independent in this sense if the operator is blindfolded and the results are recorded by an assistant. We can achieve independence either by blindfolding, or by using different operators and keeping them ignorant of each others' results.

We need now the following fundamental theorem:

A $\frac{1}{2}(1 \pm \delta)$ coin-tossing procedure can be transformed into a $\frac{1}{2}(1 \pm \delta^2)$ coin-tossing procedure by changing or not changing the effective labelling of the primary coin's faces at each trial according to the outcome of an independent trial of the same type.

The switching of labelling must take place without the operator's knowledge, or after we have recorded the actual outcome of the primary trial.

Let us refer to the outcome *heads* as H, and the outcome *tails* as T. The scheme proposed amounts to combining pairs of independent trials, treating the results HH or TT as H, and the result HT or TH as T.

We must allow the operator to be able to choose at the first and second trials respectively $\mathrm{Prob}(H) = \frac{1}{2}(1 + \delta_1)$ and $\mathrm{Prob}(H) = 1 + \frac{1}{2}(1 + \delta_2)$, where δ_1 and δ_2 have any values between $-\delta$ and $+\delta$. We then have:

$$\mathrm{Prob}(HH \; or \; TT) = \tfrac{1}{4}(1 + \delta_1)(1 + \delta_2) + \tfrac{1}{4}(1 - \delta_1)(1 - \delta_2)$$

$$= \tfrac{1}{2}(1 + \delta_1 \delta_2).$$

This expression can neither be greater than $\frac{1}{2}(1 + \delta^2)$ nor less than $\frac{1}{2}(1 - \delta^2)$. This completes the proof of the result. We lay special emphasis on the fact that the demonstration is invalid if the two trials are not independent in the sense defined above.

We may refer to the procedure involved here as the *convolution* of random sequences. It is easy to devise convolution procedures for more complex sequences. Let us suppose that at each trial, using ideal apparatus, events $A_1, A_2, \dots A_r$ can be induced to occur, each with probability $1/r$, and that with imperfect apparatus the respective probabilities behave as if

$$\mathrm{Prob}(A_i) = (1/r)(1 \pm \delta_i), \qquad i = 1, 2, \dots r.$$

Suppose, following the line of thought used in the case of coin-tossing, that a pair of independent trials is carried out. At the first we assume the operator to employ probabilities

$$\frac{1}{r} + \frac{\varepsilon_i}{r}; \qquad \Sigma\varepsilon_i = 0, \qquad |\varepsilon_i| < \delta, \qquad i = 1, ..., r,$$

and at the second probabilities

$$\frac{1}{r} + \frac{\eta_i}{r}; \qquad \Sigma\eta_i = 0, \qquad |\eta_i| < \delta, \qquad i = 1, ..., r.$$

We then have:

$$\text{Prob}(A_i, A_j) = \left(\frac{1}{r} + \frac{\varepsilon_i}{r}\right)\left(\frac{1}{r} + \frac{\eta_j}{r}\right).$$

Let us sum this over r values of the pair (i, j), choosing them so that each value of i and each value of j is represented once and once only. The terms linear in ε and η will vanish so that:

$$\Sigma\,\text{Prob}(A_i\,A_j) = \frac{1}{r} + \frac{1}{r^2}\,\Sigma\varepsilon_i\,\eta_j,$$

the summation being in each case over the r values chosen.

Now the selection of sets of pairs in this way may be represented by the laying out of a so-called Latin square with rows and columns labelled according to the numbering of the first and second members of the pair. We select from all possible pairs those corresponding to a particular labelling of cells within the square.

The maximum value of

$$\sum_1^n x_i\,y_i,$$

subject to

$$\sum_1^n x_i = \sum_1^n y_i = 0; \qquad |x_i| \leqslant X, \qquad |y_i| \leqslant Y,$$

is kXY; $k = n$, n even; $k = (n-1)$, n odd.

It follows that in the random sequence with probability parameters of the type

$$\frac{1}{r} \pm \frac{\delta}{r},$$

convolution by Latin square in pairs yields a random sequence with probability parameters

$$\frac{1}{r} \pm \frac{\delta^2}{r} \quad \text{or} \quad \frac{1}{r} \pm \frac{r-1}{r^2}\delta^2$$

according as r is even or odd. Convoluting in groups of s by means of an s-dimensional Latin structure, a condensed sequence results with what we may refer to as a maximum proportional bias of at most δ^s.

We make two final comments on the significance of the convolution procedure. In the first place, we repeat that our definition of a random event, that is to say the fundamental meaning we have attributed to the term *probability*, is absolutely necessary for the convolution procedure to be effective in the way it is. In particular, we cannot start from a sequence interpreted to be random on purely empirical evidence and use the convolution procedure to effect any necessary refinement concerning the precision of the sequence's parameters; the proof given on p. 8 fails unless we have an *a priori* criterion of uncertainty and an *a priori* criterion of probabilistic independence. Thus we may take a narrow view of the metaphysical issues raised at the beginning of this chapter, and say that a systematically constructive view of probability is forced on us if we are to provide a solution of any problem which calls for the accurate generation of random sequences. In the second place we note that we now have the means for devising randomising procedures of any required degree of refinement. When we need to invoke randomising procedures for auxiliary purposes, as for example in sampling problems, we shall be free to forget about the details of the specific randomising procedure required. We can merely assume that one sufficiently refined exists, and proceed as if no serious problem were involved.

As a simple numerical example we can consider the requirements of a randomisation procedure such as might be used for the selection of winning tickets in a lottery. Suppose, for example, that we wish to select one ticket at random from amongst a million tickets. For convenience we can take the number of tickets to be 2^{20}, and number each with 20 ordered binary digits (0 and 1). If we were able to use an ideal coin, labelling heads 0 and tails 1, then the result of a 20-fold trial would give the number of the ticket to be selected. If we can only suppose that we have, say, a $\frac{1}{2}(1 \pm 1/5)$ coin-tossing procedure, then 4-fold convolution yields a $\frac{1}{2}(1 \pm 1/625)$ procedure. Using this, we can say that the probability of getting a particular sequence of length 20 lies in the range

$$\frac{1}{2^{20}}\left(1 - \frac{1}{625}\right)^{20}, \frac{1}{2^{20}}\left(1 + \frac{1}{625}\right)^{20},$$

i.e. the probability of any particular ticket being selected is about

$$\frac{1}{2^{20}}\left(1\pm\frac{1}{30}\right).$$

We may suppose this degree of proportionate accuracy to be satisfactory, and the problem solved.

Appendix

The proof of Bernoulli's Theorem is given in most text-books of probability theory, but we include it here both for completeness and because it enables us to introduce two elementary technical devices which we shall use later. Let us suppose that in a single random trial the probability of occurrence of an event with score r is p_r. The values p_r specify a *probability distribution*. We define the *probability generating function* of the distribution as

$$f(t) = \sum_0^\infty p_r t^r, \qquad f(1) = 1,$$

and the *mean* or *expected value* of the distribution as

$$\mu = E(r) = \sum_0^\infty r p_r = f'(1) \quad \cdots \cdots \quad (1)$$

where the dash denotes differentiation with respect to t prior to putting $t = 1$.
We define the variance of the distribution as

$$\sigma^2 = E[(r-\mu)^2] = f'(1) + f''(1) - [f'(1)]^2 \quad \cdots \quad (2)$$

σ being referred to as the *standard deviation* of the probability distribution.

When the two random trials are carried out independently, the probability generating function of the sum of the separate scores is the product of the separate probability generating functions. This enables us to verify, with the help of (1) and (2), that means and variances are additive in independent trials. μ indicates the *location* of a probability distribution and σ^2 its spread. *Chebyshev's Lemma* reflects the properties thus indicated, and holds for any probability distribution:

The probability of an event outside the range $\mu \pm \alpha\sigma$ occurring is less than or equal to $1/\alpha^2$.

We prove the result as follows. We have:

$$\sigma^2 = \Sigma p_r (r - \mu)^2.$$

The value of the expression on the right-hand side is greater than the value of the same expression with the summation extended only over values of r falling outside the range $\mu \pm \alpha\sigma$. Whence

$$\sigma^2 \geqslant \alpha^2\,\sigma^2 \times [\text{Probability of an event outside } \mu \pm \alpha\sigma \text{ occurring}].$$

The result given follows immediately.

Let us consider a random trial in which the probability of an event with score 1 is p and of an event with score 0 is $(1 - p)$. We have

$$f(t) = (1 - p) + pt; \qquad \mu = p; \qquad \sigma^2 = p(1 - p).$$

In summing the results of n independent trials of the type under consideration in Bernoulli's Theorem, we have

$$\text{mean} = np, \qquad \text{variance} = np(1 - p).$$

and with r/n as the random variate instead of r we have

$$\text{mean} = p, \qquad \text{variance} = p(1 - p)/n.$$

The theorem given on p. 4 follows by substitution of these values in Chebyshev's lemma, setting $\varepsilon = 1/\alpha^2$.

The method of probability generating functions, though by no means indispensable in this or other contexts, provides a concise means for dealing with the type of sampling problem we consider later. Chebyshev's lemma is only of real use in the type of qualitative theoretical discussion involved here and in Chapter 6.

2. Information and Information Potential

There arose very early in the history of probability theory a class of problems which, contrary to the anticipations of early writers, calls for the introduction of a range of concepts beyond those involved in the discussion of games of chance. The use of randomisation in sampling procedures affords a simple example. An urn contains a large number of balls of various colours in unknown proportions. We shake it up, take out a number of balls, and seek to infer from the proportions of balls of different colours in this sample something about the proportions in the urn itself. Bernoulli himself supposed that his theorem, or a simple extension of it, might provide a sufficient theoretical basis for the procedure. This, however is certainly not true, and we require a better-defined rationale. It is strange that so simple a problem should have remained a source of controversy for over two hundred years. We can account for this as follows. In the situation envisaged, two types of uncertainty occur. The randomisation procedure, that is to say the shaking up of the urn, introduces one of these; its purpose is purely auxiliary. The other concerns the presenting uncertainty of the problem; namely our lack of knowledge of the true contents of the urn. We need to distinguish between these two forms of uncertainty, and deal with them by methods which are clearly separate in their initial stages of development. The first we associate with the term *probability* as already discussed. The second we associate with the term *information*. The contrast is that involved in the apposition of the terms *chance* and *choice* in older text-books of elementary algebra; indeed, we shall regard information theory at this stage as a simple extension of elementary combinatorial analysis. We shall discuss under its heading certain formal problems concerned with the recording and transmission of *messages*; the term *message* precisely corresponds to the old combinatorial term, *complexion*.

We shall develop the theory independently of the urn-sampling problem, although in doing so we may appear to break the thread of the main argument. There are two reasons for this. In the first place, as we shall re-emphasise in later chapters, sampling represents an especially degenerate form of information transmission; we can more easily express the fundamental

13

ideas of information theory against a wider background. In the second place, we wish to provide a basis for applications which are perhaps of greater interest than those discussed in the present book.

Symbols, Cells, Messages

We confine our attention to the most familiar type of situation in which messages are stored or recorded; as for example in the printed word, punch cards, magnetic tape, photographic plates, *and so on*. Such systems allow for a distinction between the material on which messages are imposed and the symbols of which the messages are formed. In an abstract representation of such systems, we can think of one or more cells, into each of which we can place one of two or more symbols. When we use a number of cells to record a specific message, we exclude the possibility that any of them are blank; alternatively, we regard *blank* as representing a definite symbol, open to selection from amongst the range of symbols available. When each of the cells has a definite position, as is the case in each of the examples mentioned above, we can refer to them as *localised*. The "labelling" implied by localisation is quite different from the labelling which occurs when a symbol is entered. An information store with denumerable and localised cells can be treated as the standard store, and represented by a piece of tape, divided up into successive cells and having a definite starting-point. An alternative, though less useful standard store might comprise isolated and non-localised cells to which the same symbols can be admitted. In each case we imagine the symbols to be available in specified quantities in some sort of *font*. The pervading concept is that of *distinguishability*. It is involved when we refer to a pair of symbols as either *distinct* or alternatively *identical*, or to a number of cells as *localised* or alternatively *free*; but we shall reserve the term *distinguishable* itself for messages. Thus for example, the two messages

$$\boxed{0|1|0|0|1|1} \quad \text{and} \quad \boxed{1|1|0|0|1|0}$$

each on a piece of tape with six cells, localised relative to one another, are distinguishable if the piece of tape itself is localised, and indistinguishable if it is free; 1 and 0 being distinct symbols.

In specific applications, cells may correspond to atoms or molecules and symbols to energy-levels, ionisation-levels etc.; cells to synapses and symbols to states of activation; cells to blank cages in a form and symbols to code-letters or numbers; *and so on*: but we shall confine our attention here to purely formal examples. Our primary interest now lies in evaluating the *numbers of distinguishable messages* (*N.D.M.*) which we can record, given a particular set of restrictions on the numbers and types of symbols and cells available;

that is to say, given a particular *information store*. As previously indicated, the term *complexion* is sometimes used for what we refer to here as a *distinguishable message*.

Number of Distinguishable Messages; Information Potential

We may conveniently express many standard results in the elementary theory of combinations in the present terminology. Thus we can say

(i) Given an information store comprising n localised cells with r distinct symbols each in unlimited supply, then

$$N.D.M. = r^n;$$

if the cells are free, then

$$N.D.M. = \binom{n+r-1}{r-1} = \binom{n+r-1}{n} = \frac{(n+r-1)!}{n!\,(r-1)!}.$$

(ii) Given an information store comprising n localised cells, $r \geqslant n$ distinct symbols, one only of each, then

$$N.D.M. = r^{(n)} = r!/(r-n)!;$$

if the cells are free,

$$N.D.M. = \binom{r}{n} = r!/n!\,(r-n)!.$$

(iii) Given an information store comprising n localised cells, two distinct symbols, r of one sort and $(n-r)$ of the other, then

$$N.D.M. = \binom{n}{r};$$

if the cells are free,

$$N.D.M. = 1.$$

(iv) Given an information store comprising n localised cells; k distinct symbols, r_1 of one sort, r_2 of the second sort, ... r_k of kth sort, $\sum r_i = n$, then

$$N.D.M. = \binom{n}{r_1, r_2 \ldots r_k} = n!/r_1!\,r_2!\ldots r_k!.$$

We may refer to two information stores as *independent* if the message we put in one in no way influences the type of message which we can put into the other; as, for example, it would if it reduced the supply of symbols available. If we use independent stores simultaneously, then clearly the number of distinguishable messages of the joint store is the product of the *N.D.M.* of the separate stores. This suggests that the *N.D.M.* is not likely to be the most convenient measure of the *size* of an information store; we would like the *size* of two independent information stores used jointly to be the sum of their separate *sizes*. Accordingly, we define the *Information Potential* or *Information Capacity* of a store as the logarithm of the number of distinguishable messages associated with it:

$$I.P. = \log(N.D.M.).$$

This uniquely ensures the additive property. The base to which the logarithm is taken determines the units in which the Information Potential is measured. Thus, if the logarithm is taken to base two, we say that the information potential is measured in binary units; if to base ten, in decimal units; if to base *e*, in natural units, *and so on*. We effect a transference from one unit to another by multiplying by a constant factor, determined by the formula for the change of base of logarithms:

$$\log_b a = \log_c a / \log_c b.$$

We can look at the matter as follows. If we take as standard store *binary tape*, that is to say localised cells for which there is available an unlimited supply of symbols of two types, then an information store with information potential of N binary units is equivalent to a piece of binary tape of length N; binary tape having information potential of one binary unit *per cell*. Tape using ten symbols has *I.P. per cell* of $\log_2 10$.

Frequency Constraints; the "Entropy" Function

The frequent occurrence of factorial numbers in combinatorial expressions, combined with the logarithmic definition of information potential, calls for the use of concise approximations to the logarithms of large factorial numbers. The formula associated with the first term of Stirling's expansion:

$$n! \sim \sqrt{2\pi n}\, n^n e^{-n}$$

is difficult to prove; the difficulties relate to the constant $\sqrt{2\pi}$. It is easier to derive an approximation to $\log n!$, since the constant factor is then unimportant.

If we make a graph of the function

$$y = \ln x = \log_e x$$

with x as abscissa and y as ordinate, and erect a pair of rectangles of heights $\ln r$ and $\ln (r - 1)$ on the section of the abscissa with x lying between $r - 1$ and r, for each value of r between 2 and n, we see that

$$\ln 2 + \ln 3 + \ldots + \ln (n - 1) < \int_1^n \ln x \, dx < \ln 2 + \ldots + \ln n;$$

i.e.

$$\ln (n - 1)! < [x \ln x - x]_1^n < \ln n!$$

$$\ln (n - 1)! < n \ln n - n + 1 < \ln n!. \quad . \quad . \quad . \quad . \quad (i)$$

Writing $(n - 1)$ for n, we have

$$\ln n! < (n + 1) \ln (n + 1) - n < \ln (n + 1)!. \quad . \quad . \quad . \quad (ii)$$

Whence, combining (i) and (ii), we have

$$n \ln n - n + 1 < \ln n! < (n + 1) \ln (n + 1) - n.$$

Thus, as an example, this result yields

$$361 \cdot 51 < \ln 100! < 366 \cdot 13.$$

We can infer therefore that

$$\ln n! \cong n \ln n - n$$

affords an excellent approximation for quite moderate values of n. For very large n it may even be sufficient to use

$$\log n! \cong n \log n,$$

where the base is irrelevant.

Let us consider an information store comprising n cells into which we put symbols of two types. Let us suppose that there are available r symbols of the one type and $n - r$ of the other. We have:

$$I.P. \text{ per unit cell} = \frac{1}{n} \ln \binom{n}{r} \text{ (natural units)}.$$

Let us now suppose that n, $n - r$, and r are all large. We use the first approximation given above, and after suitable reduction, we get, setting $p = r/n$, so that p is the proportion of symbols of one type used and $(1 - p)$ is the proportion of symbols of the other type:

$$I.P. \text{ per unit cell} = -p\log_2 p - (1 - p)\log_2(1 - p) \quad \text{(binary units)}.$$

We denote the expression on the right hand side by $H(p)$, which thus represents the limit to which the information potential per unit cell tends as the number of cells tends to infinity. We refer to $H(p)$ as the entropy function, and note that it is defined between $p = 0$ and $p = 1$, that $H(0) = H(1) = 0$, and that the function reaches a maximum of 1 at $p = \frac{1}{2}$.

As a simple generalisation, we note that for a very long stretch of tape using k symbols with relative frequencies $p_1, p_2, \ldots p_k, \Sigma p_i = 1$, we have:

$$I.P. \text{ per unit cell} \rightarrow -\Sigma p_i \log p_i = H(p_1, p_2 \ldots p_k).$$

The base chosen for the logarithms again determines the unit employed in measuring the information potential.

One consequence is of especial significance. If, in the case of two symbols, we take $p = \frac{1}{2}$, we have

$$I.P. \text{ per unit cell} = H(\tfrac{1}{2}) = 1 \quad \text{(binary units)}.$$

Similarly, if in the case of k symbols we take all the p's equal to $1/k$, we have

$$I.P. \text{ per unit cell} = H(1/k, \ldots, 1/k) = 1 \quad (k\text{-ary units})$$

Thus the imposition of the constraint that all symbols must be used in equal proportions has little effect on the information potential available, if coding is allowed over a sufficiently large number of cells. We use the term *coding* to denote any systematic method for representing all strings of symbols in a given set in an alternative symbolic form.

Error-correcting Codes

Let us suppose that we have an information store comprising a length n of binary tape; that is to say n localised cells into each of which we place one of two freely available symbols in recording a message. Let us further suppose that, on or after inserting a message, up to r errors can occur, each error consisting either of a 1 displacing a 0, or a 0 displacing a 1. Two related questions arise.

1. Are there any means of using such an information store so as to render the errors ineffective?

2. Can we provide a measure of the extent to which the information capacity of the store is reduced by the possibility of errors occuring?

A general solution of neither of these problems is available in the general situation under consideration; but we can provide exact solutions in special cases. Let us consider first the case where $n = 3$ and $r = 1$. The messages which the store can accommodate are:

$$000 \quad 010 \quad 100 \quad 110$$
$$001 \quad 011 \quad 101 \quad 111$$

If the message 000 is inserted, it will be read as either 000, 001, 010 or 100; and if 111 is inserted, it will be read as 011, 101, 110 or 111. Accordingly if we allow only the messages 000 and 111 to be inserted, possible errors will occasion no ambiguity at all.

The set of messages 000, 111 is a simple example of an *error-correcting code*. Its use in the circumstances we have defined results in the elimination of all errors; moreover it achieves this in the most economical way possible. Thus the information store concerned, errors subsumed, is precisely equivalent to a store consisting of one cell only of binary tape. Accordingly, the allowed possibility of error has reduced the information potential *per unit cell* from unity to one third.

The case $n = 7$, $r = 1$ also admits of complete solution. The appropriate type of code was originally devised by R. W. Hamming*. Instead of allowing any sequence of seven symbols, we may use, for example, only members of the set

$$0000000 \quad 0100101 \quad 1000011 \quad 1100110$$
$$0001111 \quad 0101010 \quad 1001100 \quad 1101001$$
$$0010110 \quad 0110011 \quad 1010101 \quad 1110000$$
$$0011001 \quad 0111100 \quad 1011010 \quad 1111111$$

If we regard the *distance* between the two *code-words* as the number of locations at which entries for the two words differ, then we can say that the code has been devised so that each code-word is at a distance of at least three from any other code-word. As in our first example, the code is perfect, since it uses up all of the 2^7 possibilities. Accordingly the error-prone information store is exactly equivalent to an error-free store comprising four cells only of binary tape. The form in which we have chosen to lay out the code makes this particularly clear; for we can regard the first four symbols of a code-word as *information symbols*, and the last three as *error-correcting symbols*. Thus in this case the possibility of error has reduced the information potential *per unit cell* from *unity* to precisely *four sevenths*.

* (1952). *Bell. Syst. Techn. J.* **31**, 504.

For larger values of n the mathematical difficulties involved in devising error-correcting codes rapidly become very great. We can, however, easily provide upper and lower bounds for the information potential per unit cell in the general case of n cells and $\leqslant r$ errors; and we can express these in a concise form when n and r are large. We use p to represent the maximum proportion of allowed errors; $p = r/n$.

Suppose that we have a perfect error-correcting code for the case of n cells and $\leqslant r$ errors. After errors have occured each code-word will be read off as one of

$$1 + \binom{n}{1} + \binom{n}{2} + \ldots + \binom{n}{np} \text{ words.}$$

We have not proved that such a code exists, but if it does, we can say that the information potential per unit cell has been reduced from unity to

$$1 - \frac{1}{n} \log_2 \left[1 + \binom{n}{1} + \binom{n}{2} + \ldots + \binom{n}{np} \right]$$

by the possibility of error; and this is the most favourable contingency.

We have (p. 18):

$$\lim_{n \to \infty} \frac{1}{n} \log_2 \binom{n}{np} = H(p).$$

Now

$$\binom{n}{np} < 1 + \binom{n}{1} + \ldots + \binom{n}{np} < (np + 1) \binom{n}{np}, \qquad p < \tfrac{1}{2}.$$

It follows, by taking the logarithm to base two of each expression in this inequality, dividing by n and taking the limit as n tends to infinity, that, as for the simpler expression,

$$\frac{1}{n} \log_2 \left[1 + \binom{n}{1} + \ldots \binom{n}{np} \right] \to H(p) \quad \text{as} \quad n \to \infty.$$

We can therefore say that, if we have a long stretch of binary tape in which the proportion of errors is either exactly equal to p or is less than or equal to p, the information potential per unit cell is reduced from unity by at least the amount $H(p)$.

We can obtain a lower bound for the reduction in information potential by using the notion of *distance* between code-words. First of all we choose, quite arbitrarily, our first code word. We then exclude from further consideration all words distant $\leqslant 2np$ from it, and choose our next code-word from amongst those which remain; we now further exclude all words distant $\leqslant 2np$ from this second code-word, and continue the process. At each stage the number of words we exclude cannot be greater than

$$1 + \binom{n}{1} + \binom{n}{2} + \ldots + \binom{n}{2np}.$$

It follows that, using this process, the number of words we can choose before the possibilities run out is at least

$$2^n \bigg/ \left[1 + \binom{n}{1} + \ldots + \binom{n}{2np} \right].$$

We can assert, then, that for very long stretches of tape the information potential per unit cell, $C(p)$ is such that

$$1 - H(2p) \leqslant C(p) \leqslant 1 - H(p).$$

Information Sources; Redundancy

We define a *discrete information source* as emitting a sequence of symbols which can be fed into a tape of indefinitely-protractable length. As characteristic examples we can think of the sequence of words generated in successive leading articles of a newspaper, of the numbers recorded in some monitoring procedure carried out at discrete intervals, *and so on*. The notion embraces any protractable sequence generated under sufficiently well-defined conditions. We consider initially two special types of information source, which we can refer to as *artificial*, insofar as they do not convey meaning, or *information* in the usual sense of the word. Firstly, consider a sequence of symbols generated according to some fully-predetermined plan such as implied, for example, in the sequences

$$010101010101\ldots.$$

$$01001000100001\ldots$$

We refer to such sources as *determinate*, and can regard their outputs as completely *compressible*, since in all such cases the infinite sequence can be replaced by a single finite message. We can reasonably say that, *per unit cell*, such a source generates virtually no *information*, or that the output of such a source is completely *redundant*.

Secondly, we consider what can be referred to as *stochastic*, or *probabilistic*, sources. A finite number of symbols is available, and at each stage the probability that a particular one is chosen depends in a specified manner on the symbols which happen to have been emitted at a given finite number of previous stages. The output of such a source is what is usually referred to as a *Markov sequence*. The simplest example comprises the sequence generated by the tossing of an ideal coin. From one point of view, this type of information source is much the same as a determinate source, since the means for generating the output of the source can be specified in a finite number of words, and we can imagine situations in which it is sufficient to communicate such a specification rather than the actual output; the latter is therefore virtually completely redundant. From another viewpoint, however, the matter is entirely different. Imagine the results of a state lottery communicated month by month to a number of newspapers. It would be absurd to allow each newspaper to generate its own set of winning numbers in its own office. In such a situation the random messages, once generated, take on a substantial meaning, so that the output of the source, though purely stochastic, has a quite definite *information-content*. The example is not altogether trivial. It affords a striking illustration of the thesis that information-content cannot be regarded as an intrinsic property of the output of an information source; and this marks a key issue in the controversies which have surrounded information theory. This discussion is allied to that on pp. 73–74 below.

Let us consider a source which, at each stage emits 1's and 0's with probabilities p and $1 - p$ respectively. We refer to this as a *binary stochastic source with parameter p*. We adopt now the second of the two viewpoints mentioned above. In a very long sequence of cells the number of 1's will almost certainly be very nearly equal to p. We infer, therefore, that it is possible, by coding over sufficiently long messages, to compress the output of the source in the ratio $H(p):1$ with negligible loss. Clearly, it is impossible to effect a greater degree of compression. Accordingly we have in $H(p)$ a meaningful and unambiguous measure of the *information-content* per unit cell of the output of the source; and $1 - H(p)$ is the appropriate measure of its *redundancy per unit cell*. Any stochastic source clearly possesses corresponding measures computable in the same way. The binary stochastic source with parameter $\frac{1}{2}$ and, more generally, any n-ary source with parameter $1/n$, are unique insofar as it is impossible to devise any means for compressing their outputs. We have no special use for stochastic sources in the present development, though, as we have indicated, they admit of formal discussion; but they require mention, since they occupy a central position in theories of information which rely on a more general notion of probability that than adopted here†.

† *Cf:* Shannon, C. E. and Weaver, W. (1949). "The Mathematical Theory of Communication", University of Illinois, 1949.

We consider now the most general class of information sources, embracing virtually all sources of practical importance. We can regard such sources as intermediate between determinate and stochastic sources, and refer to them as *indeterminate*; we may indicate the distinctions involved roughly as follows. In a communication procedure involving a sender acting as information source, and a receiver, the source is *determinate* when both sender and receiver know what the next symbol is going to be, *indeterminate* when the sender does and the receiver does not, and *stochastic* when neither do nor can. The *statistical* information theories referred to above treat every information source as stochastic, with parameters to be defined by the empirical investigation of symbol frequencies: they thus allow for unique measures of information-content and redundancy to be associated with the output of the sources which we refer to as indeterminate.

Now, clearly there is a sense in which the output of many indeterminate sources can reasonably be regarded as incorporating a degree of redundancy which is independent of the semantic function of the messages, that is to say is purely associated with the way in which symbols are being employed in rendering the required meaning; so that there arise considerations which are potentially a matter for incorporation into a theory of communication which is entirely formal. A characteristic example is the *redundancy* associated with the unequal occurence of the various letters of the alphabet in long sequences of written English. The Morse code deals with the matter by using short code-words for letters which occur frequently, and longer code-words for those which occur less frequently. We should be able to deal with such matters without making unacceptable assumptions about the nature of Probability; we do so as follows.

In respect of any system for recording or transmitting messages we may impose a restriction or constraint on the way in which the available symbols are employed. Any such constraint requires that particular messages are prohibited, and accordingly involves a definite reduction in the information potential *per unit cell* available to the operator of the system. It may be possible to devise a constraint which in no way inhibits the functioning of a particular indeterminate source. In such a case, the consequent reduction in information potential per unit cell provides a measure of the redundancy in the messages emitted from the source relative to the constraint discovered; and, in general, we can devise measures of information-content and redundancy, which are in no way absolute, but wholly depend on the ingenuity employed in surveying the effectively permanent characteristics of the sources involved.

Formalisable Features of Information Transmission

We may summarise the present chapter by relating its conclusions to the epistemological principle advanced at the beginning of Chapter1. It is at this level that we can recognise the origin of the divergence between the present approach and that of Shannon's statistical theory.

When we are concerned with systems involving the transmission or storage of information, we can focus our attention either on the means by which the transmission or storage is effected or on the information which is transmitted; that is to say either on the information channel or on the information source. We can do very little about the source, except by way of empirical investigation of its properties. We may be able to prescribe conventions to which whoever is responsible for its output must conform, or transliterate the symbols it uses by imposing coding systems upon it; but there must always be left an indeterminacy which comprises its essential feature. When our concern is with the process of transmission, we must recognise that this indeterminacy is not of our making, and therefore that we cannot make it a fully specified part of the formalism we use to describe what is going on. The channel, however, is in our hands; that is to say we are free to design or create it; and in doing so we can accurately talk of its formal specification. Accordingly, we can devise a mathematical theory of information capacity or information potential. But we cannot devise a mathematical theory of information itself. We can, indeed, think of specific messages passing through a channel we have created, or which we treat as if created, and so loosely talk of information flowing. This does not however, entitle us to regard *information* as some sort of *substance* which moves around, is transformed, which comes into being or is destroyed, in ways which can be comprehended by a formal theory.

3. Communication in the Presence of Noise

We consider in the present chapter a class of communication problems which is of special interest because it includes the problems which led to the initial development of information theory. It provides useful illustrations of the principles we have advanced in Chapter 2; and our discussion is largely concerned with demonstrating, contrary to the view of proponents of the statistical theory, that a satisfactory treatment can be given having no reference to the notion of a stochastic source, that is to say using information potential and not quantity of information as the basic concept. Furthermore, it will emerge later (Chapter 4) that certain problems, notably such as are involved in sampling theory, and usually thought of as falling wholly within the domain of probability theory, are essentially problems with the same logical status as those discussed in the present chapter. Accordingly, the discussion of the present chapter, insofar as basic principles are concerned, is of direct relevance to the work of later chapters, even though therein we shall re-examine the logical issues involved and develop what will amount to an independent line of thought.

Origins of Information Theory

It was not until the end of the eighteenth century that what has come to be universally referred to as *energy* began to emerge as a commodity in limited supply, to be manipulated by a formalism transcending the predominately geometrical orientation of seventeenth century science. In a comparable development, it was not until well into the present century that problems concerning the economical transmission of information began to call for formal concepts additional to those involved in the earlier physics of materials and of energy. The first situation of this nature concerned the allocation of radio channels, and began to attract serious attention about the year 1926.

The efficacy of a system of radio communication can be improved by either of two fundamental means; by increasing the power of the transmitter or by broadening the wave-band employed. The problem arises of providing a

25

quantative measure of the relative contributions of these two factors to the communication procedure. We think of the transmitter and the receiver as together providing a *channel* for the communication of messages. The problem is solved if we can define the *Information Capacity* of a channel, and evaluate it as a function of transmitter-power and band-width; for such a function immediately serves to specify the appropriate exchange-rate.

At the outset, a significant difficulty arises. Within the framework of classical radiation theory there is no reason why a single pulse of radiation of any given maximum size should not be used to transmit one out of an infinitely large number of messages*. Some factor other than transmitter-power and band-width must therefore enter into the required formula for channel capacity. This new factor is afforded by the random errors in messages which occur in systems of this kind as a result of disorderly activity at the particulate level in the electronic circuits involved. A realistic formal treatment of the radio communication problem becomes possible when, as in fact happened with the introduction of frequency-modulation at high frequencies, random errors come to predominate over less easily specifiable sources of message-disturbance. We refer, in general, to random errors generated independently of the generation of the message as *noise*, and we shall be lead to regard problems of communication in the presence of noise as of especial importance in Information Theory. It is at this point that probabilistic considerations begin to play an essential rôle in the theory. The principal development outlined in the previous chapter has no connections with probability theory, since, as we have already sufficiently emphasised, the notion of a stochastic source is of quite subordinate interest.

It will be convenient to employ consistently the terms *channel* and *noise* in discussing, as we proceed to do in the next section, the simplest formal problems of communication in the presence of noise. Thus, when the output of an information source is fed into an infinite tape and retrieved after the contents of cells have been subjected to errors generated by stochastic means, we shall think of the store, together with the error-producing mechanism, as providing a *channel*, the procedure followed as involving *communication*, and the error-production as *noise*. We shall use the more convenient term *capacity* when discussing what, in talking of simple information stores, we have referred to as *information potential per unit cell*.

The Binary Symmetric Channel

Let us suppose that we have a binary indeterminate information source, that is to say we envisage a situation in which an arbitrary sequence involving

---

* *Cf:* Tuller, W. S., (1949). Theoretical limits on the rate of transmission of information. *Proc. I.R.E.*, **37**, 468.

just two symbols, say 1 and 0, is fed onto what we may refer to as binary tape. Let us further suppose that at every locus

$$\text{Prob}\,(1 \to 0) = \text{Prob}\,(0 \to 1) = p,$$

where the arrow indicates change of symbol at some stage in the transmission process by random error. We refer to this as communication by way of a binary symmetric stochastic channel, or *binary symmetric channel*. If p is equal to one half, there is clearly nothing that can be done with the channel, since it is quite incapable of transmitting messages. Furthermore, we can assume that p is less than one half since, if it is greater, we merely have to interchange the rôles of the two symbols.

In using such a channel, we may mitigate the effect of errors by repeating each symbol a number of times; and we can achieve an arbitrary degree of accuracy if we are prepared to make a sufficient number of repetitions. A primitive type of coding is here involved, whereby we send only the code-words $0\,0\,0\ldots.\,0;\,1\,1\,1\ldots.\,1.$* If this is the only type of coding allowed, the capacity of the channel, *i.e.* information potential per unit cell, can only be assigned relative to a particular degree of tolerable error, the capacity relative to zero error being zero since the total elimination of error by this means involves infinite expansion of the message.

It is perhaps not immediately apparent that anything fundamentally better can be achieved. In fact the situation is quite different when other forms of coding are allowed. We proceed as follows. By taking sufficiently long code-words, we can ensure that the proportion of loci at which errors occur is as near as we like to p in any particular code-word used; and to correct such errors we can clearly, in the limit, use the type of error-correcting code discussed in Chapter 2. It follows that there exists a non-zero absolute capacity, $C(p)$, of the binary symmetric channel, and it is such that

$$1 - H(2p) \leqslant C(p) \leqslant 1 - H(p).$$

The existence of the lower bound implies that error-free communication through the binary symmetric channel is theoretically achievable without unlimited expansion of messages; and the same sort of argument would demonstrate the same to be true of a wide range of channels incorporating stochastic errors. We have established therefore, a sufficient logical basis for the solution of the type of problem mentioned at the beginning of this chapter.

---

* A minor point is perhaps worth mentioning here. When a code is employed, the sender must communicate it to the receiver, and this assuredly will take up resources of the channel. This does not, however, affect our formal definition of information capacity, which we view relative to the supposedly infinitely-protracted use of the channel.

We have, however, allowed the length of code-words used in the communication procedure to become infinitely long. This is not a practical assumption. If, as must inevitably happen, the code-words cannot exceed a given finite length, some likelihood of residual errors will inevitably remain. With this in mind, let us consider in more detail the passage to the limit which was effected in the last paragraph. A numerical example will roughly indicate what is happening. If we use words of length 4 in passing messages through a binary symmetric channel with parameter 1/10, the probability of a word being erroneously transmitted is

$$1 - (9/10)^4 = 0.344$$

If we use the Hamming code then, having regard only for the information symbols, the probability of a 4-letter word being erroneously transmitted turns out to be as low as 0.150. If, on the other hand, we go through the more wasteful procedure of repeating each letter once, we do not improve matters at all, since

$$\text{Prob} (00 \to 01 \text{ or } 11) = 0.1 \ .$$

Suppose now that we use code-words of length k. Since the channel is symmetric, the probability distribution of errors is the same for all words. Given a small number ε, we can find a number $\eta = \eta(\varepsilon)$, such that the probability of occurrence of more than $(kp + \eta)$ errors is less than ε. We can now choose an error-correcting code which corrects up to $(kp + \eta)$ errors in words of length k. The probability of a particular code-word emerging wrongly must be less than or equal to ε. We note that, provided that k is not too small, η can be chosen independently of it. Since η can be chosen so that the ratio of η to k tends to O as k tends to infinity, we have here the basis for a rigorous justification of the bounds for $C(p)$ given above. The intermediary steps used in reaching the limit are of especial interest.

We can usefully think of the totality of individual messages possible in an information store as forming the points of a space whose dimensionality is the number of cells, and for which the number of points on any co-ordinate axis is the number of types of symbol available. In the case where this is two, we can think of the distance between two points of the space, that is to say the distance between two messages, as the number of dimensions in which the co-ordinates of the two points differ. We have shown on page 19 how the devising of an error-correcting code can require the packing of "spheres" into this sort of finite space. One appropriate procedure to be employed for communicating in the presence of noise involves a simple extension of this, and any other effective procedure must be something like it. We need to pick out a number of points from the space so that each of them is surrounded by a set of points such that the total probability of either one of these or the central point being received when the central point is used as message is $\geq (1 - \varepsilon)$, and in such a way that none of the hyper-volumes touch or overlap. Now we must note that in any packing procedure where the items to be packed are of equal size, it is very unusual for a perfect packing to be devisable. In the situation examined in Chapter 2, involving non-stochastic errors this did not matter, since it could happen (and would have been quite acceptable) that certain sequences of symbols were neither used as code-words nor could emerge as received messages. When stochastic errors occur, however, it must in general happen that *any* sequence of symbols consistent with the overall set-up can actually be received. Accordingly, when the points representing code-words have been chosen, it is necessary that the points of the space which are left over should be allocated each to a particular code-word; so that the probability of erroneous transmission associated with a particular code-word may turn out to be substantially less than ε.

We may summarise our argument by saying that an effective scheme for communication in the presence of noise at the level of complexity envisaged in considering the binary symmetric channel, will involve the choice of a set of code-words and an acceptable probability of erroneous transmission for each of them; and that within this scheme every received message will be interpretable as resulting from a particular code-word or sequence of code-words. We must repeat that such a scheme allows for more latitude than occurs in the case of the error-correcting codes of Chapter 2. Since, in general, it will be impossible to associate the same probability of erroneous transmission with each code-word, it will be possible to improve, though very marginally, the communication procedure by associating code-words whose probability of erroneous transmission is especially low with messages, or segments of messages, of high semantic value. This sort of rather trivial possibility should not be allowed to divert attention from the broader and quite definite, principles underlying the formal resolution of this type of problem.

Stochastic Sources and Noisy Channels; Analysis of Received Messages

The criterion we have used for judging the efficacy of a code used for communication in the presence of noise is the maximum probability of erroneous interpretation of individual code-words. An alternative criterion which might be suggested is the *average* probability of erroneous transmission, the average being taken over all code-words employed. This will be less than the maximum probability. Thus, if we fix a particular value for the average probability of error, we can discover a larger number of code-words of a given length than if we had assigned the same value to the maximum probability of error. We are tempted to follow through this line of thought, increasing the length of code-words and allowing the average probability of error to tend to zero, and hence deriving an alternative measure for the capacity of the noisy channel involved, which will be at least as great as the capacity derived on the earlier basis. It can be demonstrated* (and the result is in any case plausible) that for the binary symmetric channel, the new capacity is in fact $1 - H(p)$, that is to say the upper bound established for what we have regarded, and continue to regard, as the true capacity. Since the averaging process takes place over messages, we are in effect following a procedure of using a stochastic source to test out the channel; thus the capacity of a noisy channel becomes the maximum rate at which stochastic source *information* can be passed through the channel with zero error.

What we have said in criticism of this use of the term *information* remains valid. But a further objection arises when it is carried into the present context. Let us think of some system devised for communication in the presence of noise such that residual errors occur. If this system has been devised on the basis of stochastic source notions, the criterion adopted for assessing the scheme's efficacy will be the overall proportion of errors in a long period of operation. From the point of view of the deviser of the scheme, this may appear to be satisfactory. From the view point of the user, who is regarded as contributing to the generation of the random sequence comprising the output of the information source fed into the channel, the matter is less satisfactory. For it may occur that a particular message he sends can be transmitted erroneously partly or wholly because of its low "probability". It will be small comfort to him that this type of experience is compensated by lower errors when other messages are sent by someone else†.

The type of problem we have examined in considering communication in the presence of noise has concerned the control of the information channel. Thus, in devising a code for the purpose of correcting stochastic errors, we are in effect replacing the given channel by another which is less noisy, and in doing so, we are expressing no interest in any particular sequence of messages which may be passed through it. It is possible, however, to contemplate a quite different class of problem. Let us suppose that we have a well-defined noisy channel, that is to say that for each symbol, word or message sent, we know the probabilities with which it will be transformed into any other symbol, word or message. We envisage the situation in which a particular message is received, and suppose that we wish to make, on the basis of this knowledge, some formal statement about the message that was sent. For example, that the sent message can be regarded, with numerically specified likelihood, as lying within a particular range of messages; that one particular

* See, for example, an article by Barnard G. A. (1956) in "Information Theory", (ed. C. Cherry).

† For a practical example, see Gilbert, E. H. (1950). A comparison of signalling alphabets, *Bell System Techn. J.* **29**.

sent message was more likely to have originated the received message than another; and so on. In short, we might seek to develop a calculus of *message-analysis* in addition to and in contradistinction to the calculus of *channel-control* which has hitherto been our chief concern.

There are, however, strong *a priori* reasons for supposing that such a calculus cannot exist. These are involved in the presumption that we are dealing with indeterminate sources. The essential characteristic of an indeterminate source is that its output cannot be completely expressed in formal terms. This being the case, it is impossible for the issues arising in message analysis to be dealt with formally. The only conceivable exception occurs when the source is stochastic, *i.e.* when, in a special sense, the output of the source *can* be formally specified. This we briefly consider.

Let us suppose that messages $A_1, A_2, \ldots \ldots A_m$ can be sent, that messages $B_1, B_2, \ldots \ldots B_n$ can be received, and that

$$\text{Prob}\,[A_i \rightarrow B_j] = \text{Prob}\,[B_j|A_i] = p_{ij},$$

where

$$\sum_j p_{ij} = 1, \text{ all } i;$$

and that the source is the particularly simple one specified by

$$\text{Prob}\,(A_i) = P_i, \qquad \Sigma P_i = 1.$$

Let us propose to assert whenever B_j is received that A_i was sent. Within the sequence of events comprising the reception of B_j, the probability we shall be right is

$$P_i p_{ij} \Big/ \sum_{i=1}^{m} P_i p_{ij},$$

so that, when B_j is received, we can regard this as the "probability" that A_i was sent. The requirement that all the "probabilities" associated with B_j add up to unity is satisfied; and they will all emerge as observable and verifiable relative frequencies in a long sequence of trials. The scheme is entirely consistent, and provides in these highly special circumstances a rational system of message-analysis.

We must note that the absence of any comparable machinery for message-analysis in the case of communication procedures employing indeterminate sources, that is to say most cases of real interest, is neither surprising nor distressing. The fact that we are able in such circumstances to devise systems of channel-control which render any formal procedure for message-analysis superfluous is sufficient to indicate that channel-control is the fundamental problem. Indeed, this is no more than an extension of the pivotal notion discussed in Chapter 2, namely the notion that the fundamental problem of Information Theory comprises the formalisation of the concept of information potential rather than of information *per se*. We may note, in accordance with our remarks at the beginning of the present chapter, that we shall in effect repeat the argument of this section in the more specialised, context of Chapter 4, and that its conclusions carry over to issues which can be discussed under the headings of sampling theory, statistical experimentation and the theory of error.

Fidelity Criteria

The rationale for communication in the presence of noise which we have outlined relates to situations in which the information to be transmitted comprises discrete symbols; furthermore, in the examples we have chosen, ultimately only a small error-probability mars the perfect reproduction of input symbols at the receiving end. A complication arises when the circumstances are such that some sort of modification of a message is acceptable, quite aside from residual stochastic errors. We can think of an ordinary photograph adequately represented in the coarse-grained version published in a newspaper, or the use of a rounding-off procedure for numerical data. We may refer to this type of situation as involving the transmission of *continuous information*, although essentially the same situation obtains when the information comprises discrete symbols, and it is acceptable to replace an individual symbol by one of a set of symbols which are in a specified sense adjacent or nearly adjacent to it.

In such circumstances, that is to say, where continuous information is transmitted through noise-free channels and suffers a limited change in the process, we can conveniently compare the totality of possible input symbols with the corresponding totality of possible output symbols by thinking of them as lying along the vertical and horizontal margins of a table or grid. We can specify the relevant characteristics of the channel by filling in the table so that for each potential input symbol we have the output symbols to which it can give rise. Alternatively, and equivalently, we can think of ourselves as possessing for each output symbol a list of the input symbols from which it could have derived. If we choose to devise such a tabulation (looked at from either viewpoint) so as to specify the minimum *desiderata* of a communication procedure, then we can refer to it as providing a set of *fidelity criteria* for the act of communication envisaged. We must then ensure that the channel chosen for the act of communication adequately accommodates the specified set of fidelity criteria. If we now add the requirement that each input symbol need only be transmitted to within the given fidelity specification with probability $1 - \varepsilon$, where ε is small, we have a suitable framework for the communication of continuous information in the presence of noise.

We must note that there is no compelling reason for looking at the matter precisely in this way. There is no reason in principle why a more intimate association between the noise-errors and fidelity criteria should not be acceptable, and such an association may accord with practical convenience. The only logical principle on which we must insist is that a coherent scheme must be envisaged within which an acceptable communication procedure appertains individually to each allowable input message.

Appendix

The logical basis of communication in the presence of noise is sufficiently illustrated by our discussion of the binary symmetric channel. For the sake of completeness we include here a brief discussion of the problem of radio noise mentioned on p. 26. We follow Shannon's specification of the problem, to which the reader is referred, and use certain mathematical results in the form given by Cassels*.

If W is the band-width, T the duration of the signal, P the maximum average transmitter power allowed, N the noise power, and if WT is large, then each signal as sent can be represented by a point lying within a sphere of radius $\sqrt{2TWP}$ in a $2TW$-dimensional Euclidean space. The effect of the noise is to deflect a signal to a distance $\sqrt{2TWN}$ that is to say onto the surface of a sphere with centre at the signal point, and radius $\sqrt{2TWN}$.

Now we can show that if $N(n, \varepsilon)$ is the number of spheres of radius ε which can be packed in a unit sphere in n-dimensional space,

$$\lim_{n \to \infty} \frac{1}{n} \log_2 N(n, \varepsilon) \geqslant \log_2 (1/\varepsilon - 1) - 1,$$

for there exists a lattice-packing of spheres for which

$$\delta_n \geqslant \frac{1}{2^{n-1}}$$

where δ_n is the density of the packing. If Λ is such a lattice, packing spheres of radius ε_1, $d(\Lambda)$ the volume of its fundamental hyperparallelepiped, and V_n is the volume of the unit hypersphere, then

$$d(\Lambda) \leqslant 2^{n-1} V_n \varepsilon_1{}^n.$$

We can place a unit hypersphere on this lattice in such a way that it will contain a number of lattice points greater than

$$V_n/d(\Lambda).$$

* See Shannon, C. E. and Weaver, W., *op. cit;* Cassels, J. W. S. (1959). "Introduction to the Geometry of Numbers", pp. 69, 247; also Rogers, C. A. (1964). "Packings and Coverings", p.3.

Any sphere of radius ε_1 with such a point as centre lies within a sphere of radius $1 + \varepsilon_1$. Thus we have

$$N(n, \varepsilon) \geqslant \frac{1}{2^{n-1} \varepsilon_1{}^n}, \qquad \text{where } \varepsilon = \frac{\varepsilon_1}{1 + \varepsilon_1}.$$

This leads directly to the first inequality above.

It follows that there exists a non-zero lower bound for the information potential C *per unit time* of the channel under consideration:

$$C \geqslant 2W \left[\log_2 \left\{ \frac{1}{\sqrt{[N/(P + N)]}} - 1 \right\} \right] - 1.$$

By an argument similar to that used in deriving the corresponding upper-bound in the case of the binary symmetric channel, we also have

$$C \leqslant W \log_2 \left(1 + \frac{P}{N} \right).$$

For high signal-to-noise ratio, C approaches the value given by the upper-bound.

4. Inverse Problems in Probability Theory

In the present chapter we resume the discussion of probability theory commenced in Chapter 1. In Chapter 1 we were primarily concerned with the underlying terms of reference of probability theory, and incidentally with the way in which probabilities are reflected in observable relative frequencies. In the present chapter we are concerned with the inverse form of this latter relationship, that is to say with modes of inference from observed frequencies to underlying probabilities. The type of problem involved is almost as old as probability theory itself, and we shall find it convenient to discuss it initially within the traditional framework. As we have already indicated, however, the logical basis of a thorough-going treatment lies in information theory; by using the language of information theory we are able to lay bare the essential cause of the controversies which have surrounded the subject. In recent years the difficulties involved have been discussed under the heading of *statistical inference*; but extraneous issues are introduced by this term, whose discussion we defer to the following two chapters.

The Nature of the Problem

When conditions for the employment of random-event probability exist, the law of large numbers allows us to identify any numerical probability with a relative frequency, provided that the number of trials can be regarded as infinitely large. In such circumstances we may immediately adopt the methods of inference commonly employed in other fields of applied mathematics. Thus, without taking any special precautions, we can use formal developments of a probabilistic theory both to make indirect measurements of entities which are defined within the theory, but which are not amenable to direct observation, and in order to derive from observation the values of unspecified parameters. Statistical physics, where the effective numbers of trials is usually very large indeed, affords many examples of such procedures; it may be noted that, significantly, many authors in this field make no distinction between probabilities and statistical frequencies.

When we cannot treat the numbers of trials involved as infinite, a difficulty arises which is peculiar to probability theory, and has occasioned a substantial part of the controversy to which the subject has proved peculiarly susceptible. In such circumstances we have to regard individual random events, and not relative frequencies, as the primary observables; the observation of a single random event is of little or no interest in itself, and has to be combined in some way with a number of other observations in order to provide a basis for the modes of inference mentioned in the last paragraph. There exists, therefore, a special type of disparity between underlying model and actual observation. We may note that we can ignore the temporal sequence involved in the generation of a random event when relative frequencies are legitimately identified with probabilities, but not otherwise; and that this turns out to be the origin of the difficulties which arise.

The earliest discussions of the subject concerned the derivation of an inverse form of Bernoulli's theorem*. It is regrettable that, from the outset, the task should have been confused with another which is quite different, namely that of delineating the domain of relevance of probability theory. People have correctly argued that, if probabilities give rise to observable relative frequencies, there must be some sense in which observed frequencies can reasonably be used to infer the values of probabilities, but have confused this with the more questionable argument that the presence of observable relative frequencies necessarily implies the existence of underlying probabilistic mechanisms which can be supposed to have given rise to them.

Since the point is crucial, and the fallacies involved so pervasive, we may mention a specific and very elementary example. Some nineteenth century text-books† give as a characteristic problem of inverse probability the assessment, on the basis of the observation of frequencies, of whether or not a coin or other apparatus involved in a game of chance is biassed. From the viewpoint adopted in the present book, this is a problem which is not amenable to investigation by the formal calculus of probability, the issues involved being those we have discussed in Chapter 1, and totally unconnected with a correct theory of inverse probability. On the other hand, we may maintain that to assess, on the basis of repeated trials, whether or not a coin is doubleheaded does involve a problem of inverse probability in the sense we regard as valid, since everything which can happen properly falls within the scope of formal probability theory. In more recent times the literature of mathematical genetics contains numerous comparable examples, in which formal statistical techniques are developed in order to appraise the validity of probabilistic models rather than to explore relationships within models established on

* See Leibniz: *Ges. Werke*, (1855) **3**, 71–97 (correspondence with J. Bernoulli, 1703–5) for the first recorded discussion.

† *E.g.* Bertrand, (1889). *Calcul des Probabilités.*

diffuse, and certainly by no means exclusively, empirical grounds. In general, we may say that a formalisable problem arises if a choice has to be made between different summetrical set-ups on the basis of a finite number of random trials.

Alternative Formulations of the Urn-sampling Problem

The simplest characteristic situation of the type to which we have drawn attention concerns the assessment of the contents of an urn containing balls of different labellings by means of random sampling. The symmetries to be adjudicated upon relate to the fact that, if labelling is ignored, the set-up remains unaltered when any two balls are interchanged. Other elementary situations to which we have referred, and which seem to involve quite different kinds of symmetries can in fact be represented by the same type of model. Thus, in place of the throwing of a cubical die, we can consider six balls each labelled with between one and six pips, and agitated in an urn prior to the withdrawal of one of them. One point, relating to our earlier remarks, is worth noting. If an urn contains a proportion of p black balls, the probability of drawing a black ball in random sampling is p. We may indeed regard it as a legitimate objective of an experimenter to assess the value of this probability, and it may well be convenient to look at the matter in this way. But we should bear in mind that our ultimate concern is with the value of the proportion from which the probability is derived, that is to say the numerical value associated with the symmetries involved.

At this point we make a sharp distinction between two ways in which any inverse problem can be formulated. Firstly, let us suppose that a number of balls has been drawn from the urn at random according to some clearly specified sampling procedure, and that we then ask what the observed properties of the sample allow us to infer about the true contents of the urn. We have, once and for all, a closed body of data in front of us, and we wish to devise a calculus for arguing back from such a closed body of data to a knowledge of the underlying structure, that is to say the contents of the urn. We refer to this as the *analytical formulation* of the inverse problem. Secondly, let us suppose that we are confronted by the urn, with its supposedly unknown contents. We enquire as to what precise plan we must draw up in employing a given sampling procedure, if we wish to acquire a particular kind of knowledge of the urn's contents. The problem in this form involves us in devising a plan of general applicability, rather than making a specific inference in a particular situation. We refer to this as the *prospective formulation* of the inverse problem.

It is remarkable that it should be difficult to detect any recognition of the possibility of an alternative to the analytic formulation in writings before about the year 1930. Later (pp. 59–60) we shall mention a possible reason why a distinction which was beginning to emerge in the succeeding decade has become blurred in more recent writings. Meanwhile, we briefly review what has become the most famous discussion of the inverse problem in its analytical formulation*. We do this for three reasons; because of its historical interest as the first systematic attempt at a solution; because it arguably provides a language into which all later attempts admit of translation; and in order to throw into relief the characteristic features of the prospective formulation, which alone we shall hold to be valid. Later, it will be of particular interest to interpret the method in the language of information theory.

The Analytical Method of Bayes

We consider Bayes's procedure as applied to the simplest urn-model situation which adequately illustrates its essential features. In doing so, we roughly follow Bayes's own presentation. Bayes's original paper, though prolix, still provides as clear an account as any of the matter. Part of the prolixity merely arises from Bayes following the Newtonian tradition and conducting his reasoning *more geometrico*; but part arises from his anxiety, unappreciated by many of his commentators, to distinguish between probabilities of the occurrence of *events* and probabilities *of the truth of statements made about events*. This will turn out to be a key issue in our discussion of the matter.

Let us suppose that we have an urn which contains two balls, each of which is known to be either black or white. There are three possibilities: both balls may be white, both may be black, or one may be white and the other black. We refer to these possibilities as *WW, BB* and *WB*. Suppose now that we draw three balls from the urn one by one, at random, and with replacement, and that each turns out to be white. We seek for a formal measure of the uncertainty or assurance we can associate with the conclusion that the urn is *WW*.

The procedure of Bayes is to introduce what have come to be called *prior probabilities*, by way of a hypothetical act of sampling preceding the actual sampling of the urn in question. Thus we think of the urn itself as drawn at random from amongst three urns, one each of the *WW-, BB-,* and *WB*-types, prior to the actual sampling taking place. The two-stage sampling will involve the occurrence of one out of 12 conceivable compound events, with the probabilities listed in the table on the *following* page.

Suppose now that, in repeatedly carrying through the double procedure, whenever we finally draw three white balls, we say the urn from which the balls have been drawn is *WW*; and that otherwise we make no statement.

* Bayes, Thomas (1763) *Phil. Trans. Roy. Soc.* **53**, 370–418.

Urn Drawn	White Balls Drawn from Urn	Probability of Compound Event
WW	0	0
WW	1	0
WW	2	0
WW	3	1/3
WB	0	1/24
WB	1	3/24
WB	2	3/24
WB	3	1/24
BB	0	1/3
BB	1	0
BB	2	0
BB	3	0

E.g. the probability of first selecting the WB urn and then drawing from that urn one white ball only in a sample of three is $\frac{1}{3} \times \frac{3}{8} = \frac{3}{24}$.

TABLE

We now have:

Probability of correct statement $= 1/3$

Probability of incorrect statement $= 1/24$

Probability of no statement $= 15/24$

From this we infer that the probability of making a correct statement, conditional upon the making of some statement is

$$\frac{1/3}{1/3 + 1/24} = 8/9;$$

and that the probability of making an incorrect statement, conditional upon the making of some statement is:

$$\frac{1/24}{1/3 + 1/24} = 1/9$$

We now interpret the values 8/9 and 1/9 as measures of the "probability", given the drawing of three white balls from a particular urn on a particular occasion (no first-stage sampling being involved) that the urn contains respectively two white balls or one white ball. We place the word *probability* in inverted commas because it no longer relates to a future event, but to a more general form of mental uncertainty.

We conclude this discussion by indicating the corresponding solution to the problem inverse to that involved in *Bernoulli's* theorem, since this was for *Bayes* himself the presenting problem*. The steps in the argument are the same as in the simpler case. Let us suppose that in a double sampling procedure there is initially a uniform probability of drawing an urn with proportion of white balls, p, from an infinite set of urns with proportions of white balls uniformly distributed over the interval 0 to 1. The compound probability that this proportion lies in the interval $p, p + \delta p$, and that r white balls are drawn in the second stage, is

$$\delta p \, p^r \, (1 - p)^{n-r} \binom{n}{r},$$

so that, given that r white balls are drawn, the "probability" that p lies between p and $p + dp$ is

$$p^r (1 - p)^{n-r} \delta p \left/ \int_0^1 p^r (1 - p)^{n-r} \, dp. \right.$$

We note that repeated integration by parts shows that

$$\int_0^1 x^a (1 - x)^b \, dx = \frac{a! \, b!}{(a + b + 1)!}, \, a \text{ and } b \text{ integers.}$$

Using this result, and treating the "*a posteriori*" probability distribution of p as one would an ordinary probability distribution, we can evaluate the "expected value" of p and its "variance" respectively as

$$\frac{r + 1}{n + 2} \cong \frac{r}{n} \quad \text{and} \quad \frac{(r + 1)(n - r + 1)}{(n + 2)^2 (n + 3)} \cong \frac{r/n (1 - r/n)}{n}.$$

The distribution of p can be shown to be approximately *Gaussian*. It follows that, for large n, the quantitive form of *Bernoulli's* theorem can be translated into a corresponding inverse form merely by interchanging p and r/n and suitably changing the form of words used. It follows, in particular, that we have obtained a justification, within Bayes' system, for the

---

* The actual words at the head of Bayes's paper are as follows: "*Given* the number of times in which an unknown event has happened and failed: *Required* the chance that the probability of its happening in a single trial lies somewhere between any two degrees of probability that can be named."

practice of attaching a standard error

$$\sqrt{\frac{r}{n}\left(1 - \frac{r}{n}\right)\Big/n}$$

to an observed relative frequency r/n.

Objections to Bayes's Procedure

There is clearly no objection to the formal derivations effected in respect of the two situations we have investigated above if the first-stage sampling is actually carried out, and if we interpret the probabilities we have placed in inverted commas as probabilities of correct assertion which have been proposed prior to the first sampling. Two objections are commonly raised to the interpretation which gives the formalism its meaning in respect of the analytical problem. Firstly, if the first-stage sampling is regarded as hypothetical, a new concept of probability has been introduced which indicates something like strength of belief, and does not admit of objective measurement. Secondly, and associated with this, an ambiguity is involved in the specification of prior probabilities, even in the simplest situations; and this inevitably leads to ambiguities in the final numerical results. Thus in the first example above we might have envisaged the WB-urn as involving two separate possibilities—that ball number one is white, ball number two black, and *vice versa*. We would then have assigned prior probabilities $\frac{1}{4}$, $\frac{1}{2}$, $\frac{1}{4}$ in place of $\frac{1}{3}$, $\frac{1}{3}$, $\frac{1}{3}$. It is easy to calculate that under the new assumption the "probability", on observing three white balls, that the urn is WW is 4/5, compared with 8/9 as obtained with the alternative distribution of prior probabilities.

These objections lose some of their force for someone who is frankly willing to admit at the outset the validity of the notion of a subjective numerical specification of probability, that is to say something akin to "betting-odds", as an extension or generalisation of random-event probability. Within such a conceptual framework we can say something such as follows in interpreting the formalism of our first example. We have drawn three balls from the urn and each one is white. If we had been prepared initially to lay odds of 1 in 3 that the urn contained just two white balls, then this fresh evidence justifies us in amending these odds to 8 in 9. The initial value $\frac{1}{3}$ is subjective, and we have assigned it as a matter of personal prejudice; but the final value, 8/9, is subjective only by virtue of the initial assignment; the calculation connecting the two is one which anyone would carry out in the same way; its value lies in its determining objectively the way in which our opinion should be changed by the appearance of fresh objective evidence.

We cannot lightly dismiss the Bayesian theory as the result of a critique whose salient points, as outlined above, are fully appreciated by many people who still regard the theory as valid. There are, however, two further objections to the theory which, being intimately concerned with its underlying purpose rather than its formal realisation, are more cogent, and to which we must now draw attention.

Firstly, we note that in endeavouring to simulate procedures of inductive reasoning which have no *prima facie* connection with *random-event probability*, the theory neglects the most characteristic feature of random events, namely what we have referred to as the temporal sequence involved in their generation. Let us suppose that, in the type of situation we have been envisaging, the experimenter chooses to *select* the evidence to which he draws our attention. Suppose that he goes through the sampling procedure of our first example, discovers a black ball in the sample he draws, withholds this information, samples the urn again, and is able to present us with the finding that three white balls have been observed. If now he requires us to say that the probability of the urn being WW is 8/9, we may complain that he has been dishonest in making a uniform assessment of prior probabilities when he was in possession of evidence indicating that one of them was certainly zero. This, however, is beside the point. The odds are already admitted to be subjective, and we might as well take it that they relate to our own state of ignorance rather than to that of the experimenter. We can get round the difficulty by asking the experimenter to provide an account of the circumstances in which the data concerned has arisen; better still, we might require him to perform the experiment before our eyes. In doing so, however, we are beginning to change the whole nature of the problem. This now begins to emerge in the prospective form. The suggestion arises that it is impossible to maintain a consistently analytical approach to the inverse problem, and that reference must somehow be made to more than the isolated data actually coming under review.

Secondly, we may draw attention again to what we have taken to be the presenting problem of inverse probability. This was not to explain how the human mind works or should work; if this had been the case we might reasonably have felt at the outset the need to define a new type of probability; and this, indeed, Bayes himself attempts. Rather, we are seeking to explain in formal terms the undeniable fact that somehow or other a random sample does give some insight into the nature of an urn's contents. In putting the matter this way there is no immediate hint at all of the need for a new type of probability. Nevertheless, it is conceivable that the matter is more subtle than one might think at first sight, and that a new type of uncertainty must be given numerical expression. We are relieved, however, of the obligation to make such an innovation if a solution is in fact available within the pre-existing framework. Such a solution exists if we formulate the inverse problem prospectively. We proceed to do this in the next section. Insofar as this formulation is successful, there is subsumed in it a sufficient indictment of the Bayesian scheme.

The Prospective Formulation; Inverse Form of Bernoulli's Theorem

The key to the solution referred to is already implicit in the lay-out of the Bayesian model prior to the distortion of the term *probability* required to accommodate the analytical problem. In that connection (p. 39) we associated random-event probabilities with the making of true or false statements, rather than with the occurrence or non-occurrence of events. This is a use of classical random-event probability with which it is difficult to find fault. As a more primitive example, let us suppose that we repeatedly toss a coin, and that when it falls we say, regardless of the *actual* outcome, that the outcome is *heads*. On any particular occasion we shall be making a statement which is true with probability $\frac{1}{2}$. This is in essence the only device we need to resolve the inverse problem in its prospective form. We extend the notion by aiming to devise schemes for generating statements of substantial significance, true with probabilities approaching acceptably near to unity.

We consider first the *prospective* problem relating to our first urn-sampling set-up. Let us take an *n*-fold sample; if all observations record a white ball, we propose to assert that the urn is *WW*; if all record a black ball, we propose to assert that it is *BB*; if some record a black ball and some a white, we propose to assert that it is *BW*. As can intermediate step in the argument, we shall examine what is the probability of correct assertion in following this plan. Clearly this will depend on the actual contents of the urn. If the urn is *BB* or *WW*, the probability is 1, if *WB*, the probability is $1 - \frac{1}{2}^{n-1}$. It follows that we can make statements which will be true with *probability* $> 1 - \varepsilon$, if we choose *n* so that

$$\frac{1}{2^{n-1}} < \varepsilon, \quad i.e. \quad n > 1 - \log_2 \varepsilon.$$

In the same way as our treatment of this problem in its analytical formulation gave us a sufficient insight into the Bayesian method, so the above discussion indicates all essential features of the prospective method. The most notable limitation forced on us is the restriction of our aim to the making of statements qualified by an *upper bound* for the probability of false assertion. This deprives the solution of the elegance which attaches to the Bayesian solution of the corresponding analytical problem. As an application of the same argument, we reconsider the *Bernoulli–Bayes* problem. We have an urn containing a very large number of balls known to be either black or white, and denote the unknown proportion of white balls by p. We are allowed to acquire information about p by sampling. Since the number of balls is very large, it does not matter whether we do so with or without replacement. Suppose we take an n-fold sample, observe r white balls, and assert that

$$\frac{r}{n} - \delta < p < \frac{r}{n} + \delta.$$

This is just the same as asserting that

$$p - \delta < \frac{r}{n} < p + \delta,$$

We can now easily work out the probability of false assertion relative to all possible values of p, using the numerical form of Bernoulli's theorem (Laplace–DeMoivre approximation). We note that this probability is greatest when $p = \frac{1}{2}$. We therefore choose a value of n such that

$$\mathrm{Prob}\left(\frac{1}{2} - \delta < \frac{r}{n} < \frac{1}{2} + \delta \,\middle|\, p = \frac{1}{2}\right) > 1 - \varepsilon,$$

where ε is small*.

We can now state an inverse form of the law of large numbers as follows: *if an event occurs with (unknown) probability p, then given any numbers ε and δ, it is possible to find an integer n(ε, δ) such that in n independent trials the probability of false assertion in saying that p lies within δ of the observed proportion of times the event occurs, is less than ε.*

This result provides a sufficient qualitative explanation of the efficacy of sampling procedures. For reasons we discuss in the next chapter, the quantitative form given to the result is arbitrary, and practical expression of the basic idea requires individual consideration of more clearly defined circumstances.

Information Theory and Inverse Probability

There exists a striking resemblance between the inverse probability problem as discussed above and the problem of communication in the presence of noise. We can explore this correspondence by expressing our first urn-model problem in the language of Chapter 3.

Let us suppose that we have a communication channel into which one of three symbols is fed at a time. The symbol A corresponds to the urn with two white balls, the symbol B to the urn with two black balls, and the symbol C to the urn with one white and one black ball. During transmission these symbols

---

* *Cf.* Feller, "Probability Theory and its Applications", p. 142.

are subject to disturbance in such a way that a sequence of A's and B's emerges as a sequence of X's and Y's. X corresponds to a white ball being drawn from the urn, and Y to a black ball. The properties of the channel are represented as follows:

$$\text{Prob}(A \to X) = 1 \quad \text{Prob}(A \to Y) = 0$$
$$\text{Prob}(B \to X) = 0 \quad \text{Prob}(B \to Y) = 1$$
$$\text{Prob}(C \to X) = \tfrac{1}{2} \quad \text{Prob}(C \to Y) = \tfrac{1}{2}$$

The information source emits a sequence of A's and B's, and the only coding procedure the sender can carry out in order to reduce the effect of noise in the channel comprises the repetition of each symbol a fixed number of times, say n. The decoding procedure which the receiver must employ is then as follows. He divides the received sequence up into batches of n letters, a starting-point having been agreed upon. If in such a batch he receives n X's, he decodes as A; if n Y's, he decodes as B; if a mixture, decodes as C.

The set-up thus described is a special case of a highly degenerate class of communication problems. It might be argued that the rôle we have attributed to the sender is quite superfluous, that a single individual is, as it were, attempting to *elicit* information *via* a noisy channel, and that indeed the term *communication* is not wholly appropriate in such a context. It is common in applied mathematics for degenerate problems to respond best at the technical level to specially devised formal methods, but to be illuminated at the conceptual level by recourse to more general considerations. This is true here. The general concept which is of especial significance, since it is particularly liable to be obscured by the special circumstances of the case, is that of the indeterminate information source.

Let us reconsider our simple urn-sampling problem in its original form, and try to put the solution we have advanced into a more concrete setting. Let us imagine that we have a large number of urns of the three types laid out in such a way that, if they were externally distinguished according to their contents some sort of picture or pattern would be visible. We now suppose that the only means of discovering this pattern is to take n-fold samples from each urn, and that we choose n so as to control the maximum probability of error, ε. What value for ε we regard as acceptable will depend on our prior anticipation of the complexity of the pattern; more accurately, on the type of complexity of which this particular pattern affords an example. Accordingly, in order to provide the basis for choosing a particular value of ε, we are led to invoke a framework of potentially protractable experience which precisely corresponds to the notion of an indeterminate information source.

We may usefully recall at this stage the general tenor of our discussion of the problem of communication in the presence of noise, and outline the argument in a form which, bearing in mind the example discussed in detail above, facilitates comparison with the problem of inverse probability.

We seek a resolution of the problem by focussing attention on properties of the channel, regarding the information sources employed as specified only by the range of messages which they can emit, and not further formalisable. By the process of deriving a suitable error-correcting code, we in effect succeed in replacing a given channel by one which is less noisy, and the scheme implied operates impartially for every allowable message. The correction of residual errors can only be affected by reference to semantic, *i.e.* non-formalisable, properties of the information source, or to extraneous knowledge. Stochastic information sources present special formal problems, albeit of little practical importance. When indeterminate information sources are employed, no formal scheme of message analysis is possible, but in the case of fully-specified stochastic sources, a consistent scheme of message analysis can be elaborated. Even in such a situation, however, channel-control remains a more significant problem than message analysis.

A parallel statement concerning problems of inverse probability runs as follows. We must view such problems prospectively and not analytically, and reject at the outset any procedure which attaches prior probabilities to the possible contents of the urn, or analogue of such; all that matters is the prior specification of all types of constitution which we can regard as possible. In the cases we have considered, the problem is resolved by fixing on a value for sample-size; and the sampling-procedure works impartially for every allowable constitution of the urn-contents. We recognise that the procedure cannot be wholly free of errors, but that in any practical situation, their correction must depend on the purpose of the enquiry insofar as it involves the building up of some sort of meaningful picture. The formal scheme of Bayes is valid if, and only if, the first-stage sampling procedure which it subsumes actually takes place; but in the absence of physically-realised prior probabilities, no formal solution of the inverse problem in its analytical formulation is possible.

Whenever we encounter a problem of inverse probability with novel features, we shall find it useful at some stage to interpret it as a problem of communication in the presence of noise. The principal usefulness of such a procedure will lie in our being forced to examine the part played in the set-up under consideration by something which corresponds to an information source.

Probability and Induction

The solution of the problem of inverse probability given in the present chapter presupposes that, without an explicit definition of a random event, no clear-cut problem exists. On the other hand, the formulation of the problem given by Bayes refers to events in general. It would appear to open up a path whereby one might hope to explore much wider issues than those implicated by the primitive doctrine of chances. The possibility was made explicit by Richard Price in the letter and appendix which accompanies Bayes's memoir, and his ideas were later taken over by Laplace in a more elegant treatment. As an illustration of the line of thought involved, we may mention the corollary of Bayes's analysis and Price's discussion, which has come to be known as Laplace's rule of succession. It purports to provide a rationale for the procedure of induction by simple enumeration, and states that if an event has occurred without fail on n successive occasions, the probability that it will occur on the next occasion is $(n + 1)/(n + 2)$.

We can obtain the result by setting up a two-stage urn-sampling procedure as follows. Assuming—what is by no means unquestionable—that such an act can be regarded as physically meaningful, we suppose an urn to be drawn at random from amongst an infinite array of urns containing proportions of white balls uniformaly distributed between 0 and 1. $(n + 1)$ balls are then successively drawn at random with replacement from the selected urn. The overall probability of all these balls being white is

$$\int_0^1 p^{n+1}\, dp = 1/(n + 2).$$

If we divide this by the probability of getting n white balls successively in an n-fold replacement sample, which is $1/(n + 1)$, we obtain the conditional probability of getting a further white ball after getting n white balls. With an interpretation which closely follows our earlier examples of the Bayesian calculus, this yields the result given above.

We have already indicated the two principal levels at which this sort of argument is open to criticism. These relate firstly to the grounds for introducing an urn model at all, and secondly to the validity of the reasoning employed in dealing with the model itself. If we reject the procedure at the first level, it is tempting to suggest that any further discussion of the matter at the second level will be wasted, so that a resolution of the urn-sampling problem which we hold to be correct will be as irrelevant to broader inductive issues as one which we hold to be incorrect. This is perhaps too drastic a conclusion. It is possible to argue that there is no single homogeneous class of formal inductive problems, and at the same time to recognise the urn-sampling problem as representative of so simple a special class, that a correct analysis may be at least obliquely relevant to others of a quite different nature.

Hume, who we may note was an almost exact contemporary of Bayes, remarks in a famous passage that there is nothing in any object, considered in itself, which can afford a reason for drawing a conclusion beyond it, a thesis whose effect is to deny all possibility of distinguishing between inductive inferences which are rational and those which are not. The result of sampling an urn containing white and black balls in unknown proportions affords an example of partial knowledge which, though deriving from a set-up which is highly schematic, is nevertheless not completely trivial. The partial nature of our implied knowledge of the contents of the urn is highly circumscribed: the uncertainties involved are under our firm control, since we introduce them by way of a randomising device of our own invention; and we are excluding *a priori* the possibility that the urn contains anything unexpected; moreover we make assumptions which on a broader view might seem to beg the question, namely that the laws relating to random events rigorously apply, and that the urn and its contents retain their identity throughout our enquiry. Yet in this situation, perhaps the least favourable to Hume's thesis that we could envisage, the conclusions of the present chapter endorse his views in their most radical expression. What, however, is of the utmost significance is that they do so without conflicting with the commonsense which Hume's more broadly-based argument has been widely held to contradict, and which in this case asserts that there must be some rational way in which the sample can be taken to tell us something about the urn-contents.

The less sceptical philosphers of induction have tended to paint a naive picture of the way in which generalisations are achieved or adjudicated upon. The investigator has before him an array of factual knowledge and may hope to relate it to an underlying theoretical structure by some formalisable procedure. The urn-sampling type of experiment affords a simple example demonstrating that this cannot be so, and that any formalisable inductive procedure must, whatever the circumstances, take into account the existence of a framework of continuing enquiry such as we have invoked when placing the problem of inverse probability in a prospective setting and utilising the notion of an information source. The conclusion emerges that we may at the same time assert that there can be no such thing as a formal inductive inference, whilst maintaining that in defined circumstances there may be such a thing as a formal regimen of inductive behaviour.

5. Random Sampling Theory

In Chapter 4 we have used simple urn models to illustrate the rationale of procedures needed for the abstract problem of distinguishing between different possible symmetrical configurations by means of the observation of limited numbers of random events. Urn-sampling, however, provides a direct representation of the randomisation procedures used in strictly practical applications, as for example in carrying out sampling surveys. We consider in the present chapter the significance for this type of application of the principles we have discussed earlier. These principles require that any assumed procedure of random selection rests on the type of physical basis discussed in Chapter 1, and they lead to the conclusion that the characteristic formal problem of sampling theory merely concerns the choice of sample size. There is implied throughout a severe limitation to the scope of sampling theory, which distinguishes it from the tangle of ideas currently associated with the term *statistical inference*. We recognise the existence of modes of inference which we can refer to as statistical, but do not regard them as formalisable.

If we view random sampling within the framework of information theory we are led to a summary resolution of difficulties which are frequently associated with problems more pretentious than those we shall be considering, and with mathematical developments more elaborate than those to which we shall need to have recourse. The need for economy in sample-size suggests that it may be helpful on occasion to carry out sampling in more than one stage. It turns out that such procedures are unlikely to be of much practical significance. But the suggestion raises a logical issue which has had a profound influence on the development of mathematical statistics during the last twenty years.

It must be emphasised that the intrinsic interest of the subject-matter of both the present chapter and the succeeding chapter—in which we examine a special application of random sampling—is slight relative to what would seem to be suggested by the vast literature of statistical theory which has grown up around the simple issues involved.

49

Statistical Inference and Frequency Probability

All writers on the history of probability theory recognise that its ultimate origins lie in the early discussion of games of chance. Many, however, attribute to later developments the introduction of a concept of probability sufficiently broad to accommodate substantial practical applications. From the viewpoint of the present book this is erroneous. We think of the crude early experiments with dice and cards as accurately indicating the essence of the concept of mathematical probability, and no less significant of its rôle in Science because two centuries elapsed before the beginnings of a revelation of their wider implications emerged in the era of Boltzmann and Mendel. A compact train of thought is involved, which precludes the application of the basic concept of probability employed substantially beyond its original terms of reference.

We may, however, distinguish another development, originating at the same time as probability theory, which from the outset has sporadically employed the same terminology, but which we must regard as virtually independent in its essential features. The post-renaissance revival of natural-istic enquiry included amongst its more novel enterprises a diffuse interest in what we would now call demographic studies. We may discover in this period the first suggestions regarding the use of statistics in the formulation of social and socio-medical policy. But the most striking and immediate interest lay in the provision of a basis for the systematic practice of life-insurance.

The first mathematicians concerned with insurance regarded life-tables, that is to say tables of age-specific mortality rates, as tables of *probabilities*, as if identifying the fundamental human uncertainty of death with the objective uncertainty of random-event probability. The idea, together with the termin-ology involved, remained in the actuarial literature for nearly two centuries. Its effect was to facilitate the provision of mechanical routines for making the statistical predictions with which insurance is essentially concerned, by deflect-ing attention from the biological issues involved and from the fallacies associated with accepting statistical data at their face value. It is a notion which modern demographers have largely rejected, so that a purely arithmetic methodology now dominates both demographic enquiry and the practise of insurance. We may associate this change of outlook with the wide knowledge which has accrued in the accumulation of mortality data over long periods of social change in widely differing societies, with an increasing awareness of the complexity of the biological processes which underlie the variability of the human life-span, and with the fact that sufficient time has passed for demo-graphers to have made predictions on the basis of observed trends in mort-ality and fertility statistics which have proved to be either wrong or of strictly limited validity. The existence of mathematical laws of mortality has long ceased to be a credible possibility. In this area at any rate, the realisation has grown up that, if we wish to make useful inferences about statistical matters, we have to proceed with a circumspection which takes into account both the empirically observed fluctuations of the type of statistical frequency involved and also related knowledge which may be expressible in neither statistical nor other quantifiable terms; we need not deny that useful predictions can be made, but we may reasonably question whether they can legitimately be mediated by any formal calculus. In inferring, for example, that a death-rate or other statistical index, will be approximately the same in one population as in another, or in one year as in the previous year, we employ a logic which is an extension of the logic of everyday life, and subject to the same sort of limitations.

In the present century what we can refer to as the actuarial method has come to be more widely employed, so that statistical methods have ceased to be concerned exclusively with commercial or economic matters, with human populations, or within human populations exclusively with the demographic characters associated with birth, death and disease. A large body of theoretical statistics has risen to accommodate such applications, explicitly or implicitly based on the frequency interpretation of probability we have noted to have been rejected by demographers. The approach involved introduces a peculiar danger. When we consider the erratic sequence of events resulting from the tossing of a coin, and ask what lies beneath the uncertainty we experience as we await the results of an individual trial, the answer is *nothing*. Everything is subsumed in the specification of the formal probabilities, and nothing of interest remains to be explored. On the other hand, if we interpret an observed statistical frequency as a probability, we obscure the fact that underlying the apparently erratic behaviour of individual events are circumstances which we have no reason to suppose are not of very genuine potential interest. In both cases the probabilistic interpretation presupposes ignorance, but in the one case it is a necessary ignorance, in the other an ignorance which is in principle remediable. In particular, if probability-based statistical methods are intro-duced into biological or sociological enquiry whose ultimate concern is with the behaviour of individual organisms, we run the risk of unthinkingly assuming a position which is such that the true objects of investigation wholly disappear from view.

We may distinguish two ways in which frequency probability is introduced into scientific discussion. The more systematic, which is largely due to Von Mises,* explicitly repudiates the existence any of *a priori* grounds for regarding a given class of events as probabilistic, and calls for extensive empirical

* See "Mathematical Theory of Probability and Statistics", 1964, and "Probability, Stat-istics and Truth", 2nd ed., 1957. For the highly relevant philosophical background of Von Mises' approach see his "Positivism: a study in human understanding", 1951.

investigation before the relevance of probability theory is regarded as established in any particular context. We have already, by implication, provided a critique of this approach in discussing the key problem of Chapter 1, which an exclusively empirical theory cannot deal with. The alternative is more protean in its applications, has a correspondingly less clearly defined rationale, and has acquired a wider currency. It admits of close association with purely subjective interpretations of probability*. Under this view:

"any body of numerical observations or of quantitative data thrown into a numerical form as frequencies may be interpreted as a random sample of some infinite hypothetical population of possible values."†

The suggestion here is not only that statistical prediction, but virtually all procedures of inductive inference should be subordinate to a logic of sampling based on probability theory. The existence of such views, especially insofar as they underlie much of the mathematical literature of sampling theory, makes it necessary to lay particular emphasis on the narrow scope of the theory of sampling advanced in the present pages. In isolating sampling theory from more diffuse issues associated with statistical prediction, we confine our attention to a special means for eliciting a special sort of information, that is to say to a special problem of indirect measurement. The fact that the sort of information involved is *statistical*, that is to say concerns relative frequencies, is wholly incidental to the logical issues which we can deal with in formal terms. It is perhaps worth noting that if it were correct for us to regard these relative frequencies as generated by stochastic means, there would be no need for subsequent sampling to be random.

Implications of Information Theory

The theories of statistical inference which have stemmed from the viewpoints to which we have just referred have been primarily concerned with the means for analysing bodies of statistical data. There have emerged well-known techniques for assessing so-called significance, and for estimating the values of statistical parameters in populations which are usually hypothetical. The common ground between our discussions and such theories is further attenuated, following our remarks in Chapter 4, by our being unable to

* *Cf.* Discussion of frequency definitions in Jeffreys, H. (1948). "Theory of Probability" 2nd ed.
† Fisher, R. A., (1925). *Proc. Camb. Phil. Soc.* **22**, 700. See also Gibbs, "Elementary Principles of Statistical Mechanics" where a similar view is expressed (p. 17): "It is in fact customary in the discussion of probabilities to describe anything which is imperfectly known as something taken at random from a great number of things which are completely described."

entertain the possibility of devising or approving such methods, even when the accumulated data results from an accredited sampling procedure which has actually been employed within a recognised finite population.

We may note that this imposes a restriction which is in no way inconsistent with the general temper of scientific enquiry. In many areas of intellectual activity a characteristic problem requires that a decision should be made, or a judgement delivered, on the basis of evidence which is palpably insufficient. In scientific investigation however, this is not a fundamental type of dilemma. The scientific worker is free to confine his attention to problems which will turn out to be soluble. Given data which is inadequate for whatever end he has in view, he will wish to set about refining or extending his methods of observation, rather than rest content with the uncertain conclusions of a quasi-judicial enquiry. This is the issue which is involved when a sampling investigation has been carried out on too small a scale.

There is a further limitation to the formal subtlety we can expect of a legitimate theory of sampling. In discussing less atypical communication problems we have examined situations where we can lay out precisely the requirements of an efficient communication procedure. In quite simple cases, however—for example that of passing messages through a binary symmetric channel—we have difficulty in deriving a solution of the coding problem involved which optimally satisfies a given criterion of economy; and we know that even if we had such a solution it would be likely to be too clumsy to be useful. Moreover we have noted that, with the trivial exception of stochastic sources, the output of any information source is necessarily transmitted with an apparent wastefulness. We must recognise that communication is an inherently wasteful procedure in which considerations of flexibility are likely to obscure the criteria afforded by the relevant theory. In the case of sampling, any formal examples we provide can only serve to indicate the underlying rationale for the rough calculations which are all that are likely to be required in practice, and to draw attention to the factors which must be taken into account in making such calculations. We may contrast this state of affairs with that obtaining in respect of analytical statistics, where the very nature of the problems posed calls for uniquely best solutions.

Single-stage Bernoulli Sampling

We have obtained (p. 44) a crude resolution of the Bernoulli sampling problem in its prospective form. Given arbitrary numbers ε and δ, we can find a number n such that, whatever the true proportion, p, of white balls in the urn, we may, as a result of taking an n-fold random sample with replacement, be able to assert that p lies within some interval of length 2δ, the probability of incorrect assertion being $\leqslant \varepsilon$. There are two qualifications to bear in mind in considering the usefulness of this type of result. In the first place it is unlikely that we shall encounter situations in which we can reasonably assume there to be no prior knowledge at all of the true value of p. We should be able to use such knowledge, where it exists, to find a smaller value of n and make the same type of statement about p. In the second place the type of statement we wish to make about p may be such that δ depends on the location of the interval concerned, and is not constant as hitherto assumed. For example, we may be interested in controlling the proportionate, rather than the absolute *error* in judging the value of p.

Such matters can be considered graphically. Let us take ε to be the proportionate area cut off by ordinates erected at distances ± 2 units from the mean of a Gaussian curve with unit standard deviation, so that ε roughly equals 1/20. For Bernoulli trials with large n we therefore have, supposing p to be known:

$$\text{Prob}\left[p - 2\sqrt{p(1-p)/n} < r/n < p + 2\sqrt{p(1-p)/n}\right] \cong 1 - \varepsilon.$$

If we plot the inequalities in the bracket on a unit square with p measured along one side and r/n along a side perpendicular to it, we get an ellipse lying with its major axis approximately along a diagonal. The normal approximation to the binomial distribution will break down at the ends of the diagonal, but as things turn out, the consequent indeterminacy of the ellipse at its extremes need not worry us. We can represent on the same diagram the range of statements we regard as acceptable outcomes of the sampling by a strip surrounding the diagonal. In the case of a fixed δ this will be a strip of uniform width $\delta\sqrt{2}$ running diagonally across the square, and the solution given above of the *Bernoulli* problem amounts to choosing n so that the ellipse is sufficiently slender to be completely embraced by this strip. When δ is taken to depend on the location of the interval in which it is said to lie, we can proceed in the same way. We describe the curves represented by the two equations

$$\frac{r}{n} = p \pm \delta(p),$$

and choose n so that the ellipse is entirely contained within them. In making a statement about p on the basis of an observed value of r, we forget about the ellipse and read off the two values of p on the boundary of the strip which correspond to r/n.

As concerns prior restrictions on p, let us confine our attention to the special case in which p is known to be less than a certain value, a. On our diagram the line $p = a$ cuts off the section of the ellipse for which $p < a$. We choose n so that the two curves representing allowed final statements together embrace this piece of the ellipse. It should be noted that the consequent sampling prescription allows statements to be made which we are supposing to be known *a priori* to be false. The probability of making such a statement is certainly less than ε; nevertheless, the existence of the possibility is likely to occasion misgivings to someone used to the analytic tradition. We are here involved with a characteristic feature of the prospective approach to inverse probability. In stating the inverse form of Bernoulli's theorem (p. 44) we have put the word *unknown* in brackets when referring to the value, p, we propose to appraise. This is because, strictly speaking, the inclusion of the word is unnecessary. The theorem holds even if p is known *a priori* to have a perfectly definite value; but again it happens that statements can be made which are known to be false. Accordingly, we must underline the point that in carrying out a sampling procedure along the lines indicated we have no right at all to say *post hoc* that a particular statement made on its basis is true with probability $1 - \varepsilon$. ε refers to the sampling procedure as a whole, and not to the individual outcome. We have here a special illustration of our earlier comments on the relationship between information channels and individual messages.

The General Problem of Single-stage Sampling

We may put these ideas into a form in which they are in principle applicable to a wider range of urn-sampling problems, in particular such as involve balls of more than two types, perhaps distinguished by numerical scoring so that our interest may lie in average values of numerical attributes rather than in proportionate numbers of balls of different types. Let us suppose that we are taking an n-fold sample from the urn, the actual value of n being as yet undetermined. We may list on the one hand all constitutions of the urn-contents which we allow to be possible, and refer to these as the *prior possibilities* or *prior hypotheses* of the sampling procedure. In the second place we can list all the possible outcomes of the single act of sampling, that is to say all the n-fold combinations of the balls of different types which are allowed by the prior hypotheses. We can now think of a table or grid, whose individual rows correspond to individual prior hypotheses, and whose individual columns correspond to individual observations. This is an extension of the procedure of p. 54. Let us now fill in the individual cells of the grid with the probabilities of particular observations arising from particular prior hypotheses. Some of these may be zero; the sum of probabilities in each row will be unity. Let us now replace a number of non-zero probabilities in each row by zero, in each case the sum of the deleted probabilities being less than or equal to ε. We refer to ε as the *uncertainty level* of the sampling procedure. Let us now imagine the sample to be drawn and a particular observation in consequence made.

We look down the column concerned, and pick out the rows in which non-zero probabilities occur. If we now state that the prior possibility actually obtaining is one of those in the set indicated, we can say that we have a procedure for generating statements which are true with probability $\leqslant \varepsilon$, *regardless of what prior hypothesis actually obtains*. We refer to the totality of statements corresponding to all possible observations as the set of *terminal statements* of the sampling procedure. Let us now suppose that *in advance* we specify a set of *acceptable* terminal statements; that is to say, we preassign a set of statements any of which, if established to be true, we would regard as justifying the performance of the trial. We now set about discovering a value of n, in particular the smallest value of n, which allows us to identify this set with the set of terminal statements reached in the type of procedure outlined above, begin free to effect this by striking out probabilities in any row of the hypothesis-observation table which add up to any value equal to or arbitrarily less than ε. We now have a complete procedure for generating statements, qualified by the uncertainty level adopted, any of which, if established, we would regard as justifying the carrying through of the sampling procedure on a particular occasion.

This looks to be a very clumsy set-up; and indeed it would be absurd to suggest that it involves a procedure to be employed in any real problem. We must bear in mind, however, that in designing any communication procedure what in fact is being done is to accommodate every one of what may be a very large number of allowed messages. If we choose to consider separately each individual possible message, *employing no short cuts*, then we reveal a complexity which frequently escapes attention.

Practical Implications

However elaborate the rationale of a sampling procedure must appear if laid down formally, what the underlying aim amounts to is simple: in a single-stage sampling investigation we need to discover a sample size which enables us to treat observed frequencies as true frequencies; what we have said earlier merely affords appropriate qualifications to this statement, and the means for translating it into formal language when necessary. It is easy, we may note, to think of cases where no finite sample size will suffice for easily stated formal aims. As a first example we may consider a situation which dominates a large segment of statistical literature. Suppose that we have a very large population in which individuals are characterised by a single numerical character, and that this character is effectively distributed normally over the population with *unknown* mean and *unknown* variance. If terminal statements comprise the placing of the mean in an interval of constant preassigned length, it is easy to see that no finite sample size suffices, and that the prospective problem is insoluble without *ad hoc* restrictions on the range of prior possibilities. This finding indicates the irrelevance of the so-called *t-distribution* to the theory of error, and one reason amongst many for questioning the validity of the more general statistical techniques developed on its basis. As a second example, let us consider Bernoulli sampling in which the set of prior possibilities embraces all values of the proportion of white balls, p, and the set of terminal statements all statements of the form $a - ka < p < a + ka$, where k is constant. Here again no finite sample suffices, since the bounds indicated will intersect every ellipse of the type discussed on p. 54 above, whatever value of n is involved.

On the other hand, if in Bernoulli sampling we are content with statements of the form $a - \delta < p < a + \delta$, δ constant, we obtain the simple result

$$\text{sample size} = \frac{1}{\delta^2}, \quad \text{when} \quad \varepsilon = 1/20.$$

That is to say, the accuracy achievable is inversely proportional to the square root of the sample size. This latter statement is, in fact, true for any value of ε. The practical significance of the result depends on our dealing with situations where values of p near to 0 or 1 are precluded *a priori*.

Text-books of sampling theory commonly justify this type of conclusion by arguments which are neither rigorous nor capable of generalisation. Characteristically there is an implied or explicit reference to a particular analytical method. A sample is supposed to have been drawn; a so-called *standard error* is attached to the consequent *estimate* of p, and n is chosen to ensure that, whatever p happens to be, this standard error does not exceed a selected value. Terminology concerned with random errors is subject to a fundamental ambiguity. The term *standard error* can refer either to the individual observations or to the procedure employed in generating observations. The *post hoc* calculation of the standard error to be associated with an individual estimate necessarily involves commitment to a particular analytical method. We may note that the way of thought lying behind a widely accepted, and seemingly harmless, practice may lead to abuse. We have to recognise that statistical enquiries, in particular those dependent upon sampling, are frequently promoted either with ill-defined aims, or with aims which cannot be achieved with the resources available. Furthermore, it may also occur that a statistical investigation is called for in order to give arbitrary decisions the semblance of objectivity rather than to produce reliable factual data. The statistician can be in an awkward position when consulted in what may purport to be a purely technical capacity, for he may be neither willing to price himself out of the market by paying too much attention to considerations of logical rectitude, nor prepared to embarrass his employers by too close a

c

perusal of their motivation. In such circumstances he may feel obliged to resolve issues such as the prior ascertainment of sample size, purely on grounds of convenience, and will appeal to analytical methods to qualify such conclusions as the inadequate resulting data indicate.

With such consideration in mind, we must be cautious about inferring the validity of a statistical method from its common employment in statistical practice. One special class of situation, however, provides a useful illustration. We have made it sufficiently clear that the logical difficulties of sampling concern the framework of potential repetition within which the individual trial is conceived. Where such repetition is to some extent realised, we may find examples which will serve to place our view of sampling methods in a useful perspective. Opinion polls intended to map voting intentions over a period of time provide a case in point. Whatever our view of their relevance to the democratic process, we may admit that the sampling procedures used are well tried, and by now fairly satisfactorily standardised. There are of course matters of detail to which we can object, but such are irrelevant in the present context. The nature of situations in which two principal candidates or parties are involved is usually such that p can be assumed a $priori$ to lie, say, between 1/5 and 4/5, so that the procedure of p. 57 can be relevant. Let us take $\varepsilon = 1/10$ and $\delta = 1/50$ (*i.e.* total width of estimating interval 4%). We then have $n = 1,600$; and this is in fact within the order of sample size commonly used in opinion polls of the type we have in mind. It turns out that with this sample size, observed results can be taken more or less at their face-value, and that fairly smooth trends emerge in appraising a moderately stable situation.

What interpretation we place on an individual result, however, is subject to the strictures we have discussed in examining the difference between "statistical" and sampling inference. When a single opinion poll is taken, in the hope of predicting the results of an election, three questions may be asked. In the first place, do the individual answers received in fact accurately correspond to the voting intentions of those questioned? In the second place, will these voting intentions, viewed statistically, change between the date of the poll and the election date? Finally, does the sample interviewed correctly reflect the state of affairs in the population as a whole? Any act of prediction in such circumstances involves all three questions; but formal considerations of probability relate only to the last

Stratification; Sequential Sampling

If we can place some limitation on the range of prior hypotheses of a sampling enquiry, it may be possible to effect a reduction in the sample size needed. An extension of this approach becomes possible if we can subdivide the population by some ancillary attribute—which in human populations might for example be age, sex or area of residence—and separately within these sub-populations assign sets of prior possibilities on the basis of prior knowledge or interpretation of the behaviour within them of the statistical frequencies in which we are interested. Such a procedure of stratification can have the further advantage of facilitating the actual procedure of selecting samples. No new principle, however, is involved, and the details of any particular application are likely to be arbitrary and uninstructive. A further possibility has a very special theoretical significance, and merits more extended discussion.

Since in Bernoulli sampling a rough knowledge of the whereabouts of the true value of p may enable us to make a reduction in sample size, we may suppose that an overall reduction might also be obtained by taking a small preliminary sample with the dual intention of providing specific information about p and of indicating an appropriate size for a second and final sample. A course which also suggests itself is that of carrying out the sampling step by step, taking single individuals at each stage until sufficient material accumulates for a terminal statement to be made belonging to the preassigned set. Let us consider what is involved in a two-stage investigation. As with single-stage sampling, we need to pre-assign a set of acceptable terminal statements and an uncertainty level; we shall also fix in advance the size of the first sample. The size, n, of the second sample will depend on the outcome of the first stage of the trial. We require a rule for choosing a value for n on the basis of the first sample, and a particular terminal statement at the end of the complete trial.

A conceptual difficulty now arises which is wholly absent in the case of single-stage sampling. Let us suppose that we wish to compare two sets of rules in order to find out which is the better, in the sense of leading to the smaller value of n. We might think of trying to work out, relative to each prior possibility separately, the probability distribution of n in respect of each suggested scheme. We at once encounter two dilemmas. In the first place, how do we judge one of these probability distributions to be more favourable than another, save by arbitrarily focussing attention on a single parameter, say the mean? In the second place, what do we do if, as is likely, one scheme appears better than the other relative to each member of one set of prior possibilities and worse relative to each member of another?

It appears that we can discover the optimum procedure to follow only if we already know with supererogatory precision what it is the object of the investigation itself to determine. We need not, however, regard this as paradoxical. The set of prior possibilities adopted represents a public ignorance, not the private ignorance of the investigator, and it is this public ignorance which has to be resolved. We are not obliged, in viewing formally any measuring procedure, to require that the observer should be formally accountable for the methods he uses, only that the actual result attained should possess objective validity. In the present situation, a perfectly acceptable procedure might be for the investigator to guess the true prior hypothesis, that is to say the true contents of the urn, and discover the rule which minimises the expected sample size.

It is important to note that if we reject this sort of approach, and call for a completely formal development of the principles of sequential sampling, a large part of the argument of the present book crumbles. For we should habe to formalise the prior uncertainties of a sampling procedure in terms of what would essentially be Bayesian probabilities. As a consequence, we should lose the sharp distinction between prospective and analytical procedures, and the whole relevance of the information approach*.

Having recognised the significance of sequential sampling in theoretical discussion, we must note that from the present viewpoint, its practical uses are likely to be very limited. We can contemplate an idealisation of the Bernoulli sampling situation in which the prior possibilities which must be entertained comprise all values of p between 0 and 1; but it is difficult to think of a practical situation in which this could reasonably be supposed to obtain. Any limitation we can place on the range of prior possibilities will limit the degree of economy achievable by a sequential procedure. Furthermore, we should note that the eventual sample size in sequential sampling can never be *guaranteed* to be less than that required for single stage sampling; that the mathematical difficulties involved in devising a sequential scheme, even with the simplest specification, are very considerable; and finally, that the unpredictable saving associated with a possible reduction in sample size is likely to be outweighed by the mere inconvenience of sampling in more than one stage.

We may conclude this chapter by noting that the beginnings of a systematic prospective approach to sampling theory occurred with J. Neyman's theory of Confidence Intervals and R. A. Fisher's theory of Fiducial Probability†. We have no need to analyse these theories in detail, since they were developed by their authors to fall in with the traditional analytical orientation. But we may observe that, in effect, each theory recognises in its initial stages the hypothesis-observation table of p. 55 above and the formal possibility it affords for making statements with controllable probability of error. The theory of confidence intervals encounters an irresolvable difficulty in seeking

---

* For a development of this alternative, see Abraham Wald, (1950). *Statistical Decision Functions*, N.Y., on which much later work has depended.

† See respectively *Phil. Trans. Roy. Soc. A*, **236**, 1937, 333 and *Proc. Roy. Soc. A*, **139**, 193, 343.

to devise objective rejection criteria for the hypothesis-observation table. The fiducial theory recognises the nature of this difficulty, but deals with it by placing an intolerable restriction on the range of formal situations which the theory can accommodate.

6. Statistical Experimentation

We consider in the present chapter the subject of statistical experimentation; more precisely, the use of randomisation as an ancillary device in experimental procedures relating to variable, notably biological, material. It turns out that this involves the assessment of statistical frequencies. The latter do not, however, enter into the formulation of the underlying problem.

Just as we were able to discuss the fundamental issues of sampling theory more freely against the background of information theory, so in the present chapter we shall be able to discuss statistical experimentation better against the same background and in the light of the general conclusions of our earlier discussion of sampling theory. We need a single new concept, concerning what we shall refer to as *incommensurable attributes*. We shall maintain that a statistical experiment serves to explore the statistical distribution of a pair of such attributes, and that therein lies the key to a correct appraisal of its role in biological and sociological fields, and an explanation for the feeling common to many scientists that statistical methods have something special to offer beyond their more prosaic applications. We shall conclude that this anticipation has small likelihood of significant realisation.

Incommensurable Attributes

We may speak of an attribute ascribable to each member of a class of organisms when a communicable criterion is available for dividing the class into two or more clear-cut, or approximately clear-cut, sub-classes. Commonly, we think, for example in considering human beings, of attributes, like eye-colour, weight or blood-group whose assessment involves no serious interference with the organism, which can be assessed straight away, and which remain stable over a period of time. There is no need, however, to restrict ourselves in this way. As an example, we might think of classifying a group of people on the basis of potential weight increase over a given period of time in a given diet. Here, in order to assess the attribute in respect of an individual subject as he exists at a particular time, we must submit him to a prescribed regimen. Thus we envisage the attribute as pertaining to the individual at a particular point in time, but do not regard it as actually assessable then.

A special situation arises when two attributes of this type are thought of simultaneously. Suppose, for example, that concurrently with the classification by weight increase proposed above, we wish to consider the alternative classification by weight increase under a different dietary regimen. In the case of any particular individual, we shall be unable to make a direct assessment of both attributes, since the act of assessing one will destroy the possibility of assessing the other. We may refer to such pairs of attributes as non-coascertainable, or *incommensurable*. As the example we have used indicates, Medicine affords a particularly striking class of incompatible characters. When two alternative therapeutic measures present themselves, then from the patient's viewpoint we would wish to be able to envisage in his individual case their separate outcomes, and from the experimental point of view we might like to be able to assess succesively their outcomes having restored the *status quo* in between. If we look at the matter against the background of this type of situation, then in the simplest case we shall think of each member of a pair of incommensurabe characters as involving a binary classification of the organisms concerned into potentially *curable* and potentially *non-curable*.

In considering such a pair of incommensurable attributes, we can usefully represent each individual by a coin. One side of each coin is, say, red, and exclusively concerns the response to treatment A; the other blue, and concerns the response to treatment B. If each attribute involves a simple binary classification, we can represent the two alternatives by *plus* (potentially curable) or *minus* (potentially non-curable) signs on the appropriate sides of the coins. The incommensurability of the two characters is represented by the requirement that we may only inspect one side of any particular coin. The possibility arises of attempting to resolve the dilemma of incommensurability, given a number of coins of the four different types (which we can denote by $+ +, + -, - +, - -$) in unknown proportions, by taking some *at random* and inspecting the red side, another group *at random* and inspecting the blue side. Such a procedure, we may suppose, will yield information about the overall proportions of coins of the four types. We may think of this set up as the model representation of an especially simple statistical experiment.

According to this definition, any statistical experiment involves direct application of a sampling procedure. In considering ordinary attributes we have supposed the sampling to be with replacement, even though in actual applications non-replacement sampling would most likely be more convenient. For samples from very large populations the two are nearly equivalent, and, in any case, in respect of ordinary attributes, replacement sampling is allowable in principle. Quite the opposite is true when we come to consider the sampling of incommensurable attributes. Here we may well envisage a situation in which the sampling procedure requires that the whole experimental group is divided up at random into the two classes within which the incommensurable attributes are to be separately assessed. We have particular need, therefore, to bear in mind the type of mathematical procedure which deals with non-replacement sampling.

Non-replacement Sampling; One Simple Attribute

The method of probability generating functions is especially suited to problems of non-replacement sampling. To illustrate its use, let us consider first the derivation of the probabilities associated with the random partition of the contents of an urn containing balls of two types only.

Let there be N balls in the urn, of which Np are white and Nq are black, $p + q = 1$. We draw n balls from the urn at random without replacement. Let the number of white balls drawn be r, the number of black balls drawn $n - r$.

There are $\binom{N}{n}$ distinguishable ways of drawing n balls from amongst N, $\binom{Np}{r}$ ways of selecting the r white balls from amongst the Np white balls, and $\binom{Nq}{n-r}$ ways of selecting $n - r$ black balls from amongst Nq black balls.

The probability of getting r white balls in an n-fold sample is therefore

$$p(r) = \binom{Np}{r} \binom{Nq}{n-r} \bigg/ \binom{N}{n}.$$

The probability generating function for r is therefore:

$$\sum_r p(r)t^r = f(t) = \frac{1}{\binom{N}{n}} \sum \binom{Np}{r} \binom{Nq}{n-r} t^r.$$

This is the coefficient of S^n in

$$\frac{1}{\binom{N}{n}} (1 + tS)^{Np} (1 + S)^{Nq}.$$

We have:

$$f'(t) = \text{coefficient of } S^n \text{ in } \frac{1}{\binom{N}{n}} Np\, S(1 + tS)^{Np-1} (1 + S)^{Nq},$$

$$f''(t) = \text{coefficient of } S^n \text{ in } \frac{1}{\binom{N}{n}} Np\,(Np - 1)S^2 (1 + tS)^{Np-2} (1 + S)^{Nq}.$$

We can obtain the mean and the variance of the probability distribution of r by putting $t = 1$, and using the formulae of p. 11;

$$\text{mean} = np, \qquad \text{variance} = \frac{N - n}{N - 1} npq;$$

and we can compare these with the corresponding values for replacement sampling:

$$\text{mean} = np, \qquad \text{variance} = npq.$$

When N is very large relative to n, the two variances are approximately equal, as one would expect.

Non-replacement Sampling: Two Incommensurable Attributes

As a second example, let us consider the simplest situation involving a pair of incommensurable attributes. The procedure we need to follow is parallel to that employed above, but more complicated; the probability generating function approach is almost indispensable. We can keep the underlying object of the calculations in mind by talking of the urn-coin set-up of p. 63 in terms of *treatments* and *responses* to treatments.

Let us suppose we have N individuals hypothetically classifiable into four classes as in Table (I). When $\delta = 0$ we have the situation in which *Treatment A* is regarded as *no treatment* and *Treatment B* can be assumed to have no "side-effects" reflected by the cure criterion.

TABLE I

$\alpha + \beta + \gamma + \delta = N$		Potential Response to Treatment B	
		$+$	$-$
Potential Response to Treatment A	$+$	α	δ
	$-$	β	γ

We partition them at random into two groups, and assess the responses to *Treatment A* in one group and the responses to *Treatment B* in the other group.

Let us suppose the results of this are as follows:

TABLE II

$m+n=N$	Response		Total
	+	−	
Treatment A	a	$m-a$	m
Treatment B	b	$n-b$	n

Corresponding to the two characters involved, we have two random variables, a and b, to consider; alternatively the two *cure-rates* a/m and b/n. It is reasonable to think that we shall be particularly interested in the difference between the two cure-rates. Accordingly we work out in the first place the joint probability of a and b; thence we can derive properties of the probability distribution of $(b/n - a/m)$. The probability distributions of a and b are, of course, not independent; the variance of $(b/n - a/m)$ will depend on the expected value of the product of a and b, which is in general not zero.

We draw m subjects at random from the group in Table I. Let us suppose that of these respectively $u, v, m - u - v - w, w$ are of the types $\alpha, \beta, \gamma, \delta$. The probability of this occurence is:

$$\frac{1}{\binom{N}{m}} \binom{\alpha}{u} \binom{\beta}{v} \binom{\gamma}{m - u - v - w} \binom{\delta}{w}$$

From Table II we have:

$$a = u + w, \qquad b = \alpha + \beta - u - v,$$

so that

$$\text{Prob}\,(a, b) = \frac{1}{\binom{N}{m}} \Sigma \binom{\alpha}{u} \binom{\beta}{v} \binom{\gamma}{m - u - v - w} \binom{\delta}{w},$$

where the summation is over all values of u, v, w consistent with $u + w = a$, $\alpha + \beta - u - v = b$ and the meaningfulness of the combinatorial bracket expressions.

We write down the joint probability generating function of a and b, attaching to them respectively the dummy variables s and t:

$$\frac{1}{\binom{N}{m}} \Sigma \binom{\alpha}{u} \binom{\beta}{v} \binom{\gamma}{m - u - v - w} \binom{\delta}{w} s^{u+w}\, t^{\alpha+\beta-u-v},$$

the summation being now over all values of u, v, w merely consistent with the meaningfulness of the combinatorial brackets.

This expression is equal to the coefficient of S^m in

$$\psi(s, t) = \frac{t^{\alpha+\beta}}{\binom{N}{m}} (1 + Sst^{-1})^\alpha (1 + St^{-1})^\beta (1 + S)^\gamma (1 + Ss)^\delta$$

$$= \frac{1}{\binom{N}{m}} (t + Ss)^\alpha (t + S)^\beta (1 + S)^\gamma (1 + Ss)^\delta.$$

This expression contains everything we need to know about the probability distribution of (a, b). Let us calculate the expected value and variance of $(a/m - b/n)$.

We can obtain the means and the variances of (a/m) and (b/n), and the mean of their difference directly from the results of p. 65 without using ψ:

$$E\left(\frac{a}{m}\right) = \frac{\alpha + \delta}{N}; E\left(\frac{b}{n}\right) = \frac{\alpha + \beta}{N}; E\left(\frac{b}{n} - \frac{a}{m}\right) = \frac{\beta - \delta}{N}.$$

$$\text{var}\left(\frac{a}{m}\right) = \frac{n}{(N-1)m} \frac{\alpha + \delta}{N} \cdot \frac{\beta + \gamma}{N},$$

$$\text{var}\left(\frac{b}{n}\right) = \frac{m}{(N-1)n} \frac{\alpha + \beta}{N} \cdot \frac{\gamma + \delta}{N}.$$

Given any random variables x, y, we have

$$\text{var}(x - y) = \text{var}(x) + \text{var}(y) - 2E(xy) + 2\bar{x}\bar{y}$$

since $\text{var}(x) = E(x - \bar{x})^2$ and the expected value of any sum or difference is the sum or difference of the separate expected values, regardless of whether the variates are independent or not.

All we need therefore, is the expected value of ab. This is the coefficient of S^m in

$$\left(\frac{\partial^2 \psi}{\partial s \, \partial t}\right)_{s=1, t=1}$$

Calculation yields

$$E(ab) = \frac{mn}{N(N-1)} [(\alpha + \delta)(\alpha + \beta) - \alpha],$$

and substitution of the various results we have obtained and appropriate reduction yields the final results:

$$\text{variance} \left(\frac{b}{n} - \frac{a}{m} \right) = \frac{n(\alpha + \delta)(\beta + \gamma)}{m(N-1)N^2} + \frac{m(\alpha + \beta)(\gamma + \delta)}{n(N-1)N^2}$$

$$- \frac{2(\alpha + \delta)(\alpha + \beta)}{N^2(N-1)} + \frac{2\alpha}{N(N-1)} \, .$$

When $m = n$, this reduces to

$$\frac{1}{N^2(N-1)} [(\alpha + \gamma)(\beta + \delta) + 4\alpha\gamma] = \frac{N(\alpha + \gamma) - (\alpha - \gamma)^2}{N^2(N-1)} \, .$$

An alternative, but much more laborious, procedure would involve noting that the probability generating function of $(a/m - b/n)$ is the coefficient of S^m in

$$\psi \left(r^{1/m}, r^{-1/n} \right).$$

Statistical Resolution of the Therapeutic Trial Dilemma

The result we have just obtained enables us to indicate in precise terms the nature of the resolution which is achievable by the use of randomisation, of the dilemma presented by the existence of incommensurable attributes. We shall find it convenient, here and later, to speak in terms of medical applications, and think of the presenting difficulty as the *therapeutic trial* dilemma, though anything we say has relevance within a wider context.

If in Table I we think of *Treatment A* as the standard, or no, treatment, and of *Treatment B* as the new treatment whose efficacy is to be assayed, then the improvement in the cure-rate achievable by submitting all subjects to *Treatment B* rather than to *Treatment A* is

$$\frac{\alpha + \beta}{N} - \frac{\alpha + \delta}{N} = \frac{\beta - \delta}{N} \, ,$$

Clearly this is what we are especially interested in. When we speak here of a *cure-rate*, we mean no more that the *de facto* proportion of subjects cured according to a preassigned criterion, and there is no suggestion that a treatment has associated with it an ideal cure-rate ascribable to an abstract or hypothetical statistical population.

We have demonstrated that, on making a random partition of the subjects into two equal groups, and submitting one group to *Treatment A* and the other to *Treatment B* with results as in *Table (II)*, then the probability distribution of

$$x = \frac{(b-a)}{n}, \qquad \left(m = n = \frac{N}{2}\right)$$

is such that

$$\text{mean}\,(x) = \frac{\beta - \delta}{N}\,; \quad \text{var}\,(x) = \frac{N\,(\alpha + \gamma) - (\alpha - \gamma)^2}{N^2\,(N-1)},$$

The latter attains a maximum such that

$$\max\,[\text{var}\,(x)] = \frac{1}{N-1}\ \text{when}\ \alpha = \gamma = N/2,\, \beta = \delta = 0.$$

Accordingly, by Chebyshev's lemma, we can say that

$$\text{Prob}\,\left\{\frac{\beta - \delta}{N} - \frac{\alpha}{\sqrt{N-1}} < x < \frac{\beta - \delta}{N} + \frac{\alpha}{\sqrt{N-1}}\right\} \leqslant 1/\alpha^2$$

whatever the true values of α, β, γ, δ; and that

$$\text{Prob correct assertion}\,\left\{x - \frac{\alpha}{\sqrt{N-1}} < \frac{\beta - \delta}{N} < x + \frac{\alpha}{\sqrt{N-1}}\right\} \leqslant 1/\alpha^2.$$

It follows that if we are willing to make statements putting the difference between cure-rates in an interval of length l with preassigned probability $\leqslant \varepsilon$ of being wrong, we can certainly achieve this if we are able to work with an experimental group such that

$$N = 1 + \frac{4}{\varepsilon l^2}\,.$$

This is intended purely as a rigorous proof of the qualitative proposition, which the reader may be willing to accept intuitively, that a statistical resolution of the therapeutic trial dilemma is possible. Providing a formal proof merely compels us to make the terms of the resolution explicit. The proof applies *a fortiori* to situations in which *a priori* restrictions can be placed on the values of α/N, β/N, γ/N, δ/N, with N anticipated to be large; to situations where the prescribed interval-length, though everywhere greater than zero, is not uniform; and to cases in which ε, though it would naturally be expected to be taken to be always small, is not strictly independent of α, β, γ, δ. It can easily be extended to cases where N and $2n$ are not assumed to be equal, and to cases where responses to treatments fall into more than two categories, subject to a system of scoring.

Limitations of Statistical Experimentation

Since we conceive statistical experimentation to be a special case of random sampling, much of our discussion in Chapter 5 carries over directly, in particular our comments deriving from the analysis of sampling procedures in terms of prior hypotheses and terminal statements, and our brief discussion of stratification and sequential sampling. As there, the *ad hoc* nature of any realistic assignment of sets of prior hypotheses and sets of terminal statements inhibits the provision of illustrative examples. By viewing throughout random sampling within the framework of the information concepts discussed earlier, we have ensured, however, that the formal issues raised by our analysis are in fact those of the most compelling significance. Thus we are unable to thereby concern ourselves with problems involving the analysis of the results of statistical experiments in circumstances deemed to be appropriate on incompletely specified grounds. Instead, our attention being drawn to the qualities of a special type of information channel and therefore to the qualities of a special type of information. We are led to focus our attention on the scope and limitations of statistical experimentation and its relationship to alternative methods of procedure. Here several of the issues discussed in Chapter 5 assume a special interest.

In the first place, we must emphasise that a statistical experiment which is to result in a statement taken from amongst a set of statements each of acceptable precision inevitably requires a large number of trial subjects. It is difficult, indeed, to envisage a situation involving the simple *cure/non-cure* type of response, in which less than, say, five hundred trial subjects in each treatment group would be called for; and this is a very cautious assessment of the situation. This is an important conclusion of our analysis, and requires especial emphasis, since a contrary view is frequently expressed*.

Our remaining remarks concern the nature of the terminal statements a statistical experiment allows. In discussing the sampling of simple attributes, we have distinguished between the formalisable probabilistic inferences relating the sample to the population sampled, and the non-formalisable statistical inferences involved in relating conclusions to other populations. When incommensurable attributes come into consideration the distinction is of especial significance. In the case of simple attributes it may well happen that the whole subject of a sampling enquiry concerns properties of the population actually sampled. Whereas in the statistical experiment, we have a type of situation in which it would be difficult to ascribe any meaning at all to the

---

* See Fisher, R. A., "The Design of Experiments", and many derivative works by later authors dealing with small sample analytical methods.

sampling procedure were it not anticipated that the results would have some relevance, however oblique, outside the population sampled. This is true whether we envisage partitioning the whole population at random, or taking two non-replacement samples of total size less than that of the population; in either case some mode of informal statistical inference must mediate the subsequent usefulness of the result.

Let us think of the therapeutic trial situation of our earlier discussion, and ask what a terminal statement resulting from a single experiment means. Suppose that the experiment yields a value for $(\beta - \delta)/N$ which is unequivocally greater than zero. In what sense can we say that Treatment B has been demonstrated to be better than Treatment A? We may have no reason to assume that $-+$ individuals do not exist, that is to say $\delta = 0$; accordingly we certainly cannot say that the subjects involved would have all benefited by being submitted to Treatment B rather than Treatment A. The most we can say is that Treatment B has been demonstrated to be *statistically* better than Treatment A for the group which has come under surveillance (as it existed at the outset of the experiment) and subject to the assumed relevance and validity of the cure-criterion adopted. There are three distinct qualifications here, which the reader may usefully consider as they relate to such concrete situations as may come to mind.

We may question therefore the view that in a characteristic field of application such as medicine the statistical trial affords a court of appeal which has in any sense a unique or final authority. The statistical trial provides a special type of information, whose primary limitation lies in its relating to the group, rather than to the individual organism; whether or not there are circumstances in which such information can be usefully assimilated with other sorts of information in formulating an insight into individual behaviour is an issue which we cannot judge in the abstract. We can, however, say that the performance of a statistical trial is necessarily attended by great difficulties, not the least of these being the large number of individuals which must take part; and that if we had to rely on the statistical method for the validification of all advances in a field such as therapeutics the prospect would be bleak*. For the same reasons we may question any suggestion that the introduction into the social sciences of methods of statistical experimentation can accord them the status of parity or of comparability with the experimental sciences. Indeed the introduction of purposive statistical experimentation advances us very little beyond the familiar situation which obtains when adventitious statistical data is used to substantiate preconceived views of causal relationships. We can regard such inferences as based on the view that nature (or society) occasion-

---

* For old, but still relevant, discussion of the role of statistics in Medicine, see Claude Bernard's *Introduction to the Study of Experimental Medicine* and Almroth Wright's *Studies in Immunisation*, 2nd Series.

ally performs statistical experiments which await our systematic appraisal. In our analysis of the statistical experiment purposively conceived and executed we have a firm delineation of one set of limitations to the validity of such reasoning. These limitations are more fundamental than those deriving from the mere recognition that Nature usually performs her experiments in a clumsy and inefficient manner, and in themselves are wholly concordant with the suspicion that often, and rightly, attaches to reasoning which rests on this type of evidence.

In conclusion we must note that the discussion of the present chapter exclusively concerns situations in which the existence of the relevant incommensurable attributes can be presupposed. Let us consider, an as example of another type of situation, the trial of a vaccine in a small community in respect of an infectious disease. If we proceed by putting half of the population at random into the treated group and half into the untreated group, then the results of the trial will be summarised in a table like Table II of p. 66. But, since the responses of individuals participating will not be independent, we cannot interpret these results against the background of a set-up such as is represented in Table I, and it will be impossible to regard the experiment as subsuming any clearly-defined measurement procedure.

The ensuing and final section of the present chapter is independent of the argument presented above.

Subsidiary Notions of Experimental Randomisation

The intention of the present chapter has been to delineate what turns out to be the very limited scope of randomisation, *sensu stricto*, in procedures of biological experimentation. As such, what we have said depends entirely on the discussion of Chapter 1 and the rigorous definition of the notions of probability and randomness contained therein. We note here a possible source of confusion associated with a use of the term *random* which is fundamentally quite distinct from what we have chosen to adopt.

Let us consider the following situation. In a physiological experiment the experimenter proposes repeatedly to apply two stimuli in some sort of succession, recording the immediate responses in the hope of establishing a causal relationship between stimulus and response. A first suggestion might be that he should first apply one stimulus, then the other, and so on in strict alternation. We might object to this that a seemingly consequent alternation of responses might merely reflect on inherent rhythm of the organism involved, unrelated in fact to the pattern of stimuli employed. In order to accommodate this sort of objection the experimenter considers administering the alternative stimuli at random, by tossing a coin. In doing so, however, he runs a risk. By chance the coin might come down heads, say, on every one of the presumptively small number of occasions involved, and so make the experiment quite pointless; or it might come down alternately heads and tails, and so produce the very type of regularity it was the whole object of the exercise to avoid. To circumvent this difficulty the experimenter tentatively proposes to repeat the randomisation procedure if a sequence arises which is clearly inacceptable, having in mind that a logical inconsistency may be involved. We may, however, easily see that the procedure is at root justifiable; and for reasons which serve to divorce the considerations discussed entirely from the notions of probabilistic experimentation entertained in the present chapter.

This we may do if we recognise at the outset that it is not the object of the experimenter in the circumstances outlined to generate a sequence absolutely random in the probabilistic sense, but to devise for very specific purposes of his own a sequence which is disorderly in a purely static sense, possessing the properties of what we may refer to as a *static chaos*. The problem of devising a static chaos is purely mathematical. To devise a sequence with the attributes of a static chaos we must first lay down *a priori* criteria for the types of regularity or pattern we regard as relevant to the problem in hand. We then write out all possible sequences of the required length, and strike out those which exhibit the types of regularity we have specified. Those that remain we may regard as 'random', and in the circumstances to which we have referred the experimenter is free to choose any of these in laying out his scheme of alternative stimuli*. We can now easily explain the rôle of the coin-tossing solution. If we increase the number of terms in the sequence envisaged, the number of possible sequences increases at a far greater rate than the number of patterns of regularity specifiable on any set of reasonably plausible criteria, so that we can anticipate that the proportion of sequences excluded in the deterministic procedure we have outlined gets smaller and smaller. Accordingly we can say that, provided the sequence involved is sufficiently long, the probability of generating a static chaos by using a dynamical procedure of physical randomisation is nearly unity. The experimenter is using the coin-tossing procedure as an analogue device for solving a purely computational problem. The notion of rejecting inacceptable (and unlikely) sequences satisfactorily fits into such a scheme.

What we have said here underlines what we have earlier perhaps sufficiently emphasised; namely that the notions of probability and randomness are to be interpreted within a framework in which a definite temporal succession is always implied. We have referred to *random events*. This is only correct in a qualified sense. In one sense a random event ceases to be random as soon as it occurs. There have been published in the course of the last forty or fifty years a number of sets of tables of so-called *random numbers*. The means used for

*Cf. section 50 of K. R. Popper's "Logic of Research", where a similar procedure is put forward in the context of a general discussion of probability theory.

generating these have been various, and in some cases bizarre. There is little point, however, in attempting to adjudicate between them if we hold that a number generated as random by the most perfectly acceptable means ceases to be random as soon as it is committed to paper. We can, however, admit there to be a kernel of sense in the notion of a table of random numbers; but we must recognize that this is concerned only with the type of static chaos or static randomness we have discussed above, and if useful at all, the utility of a table of random numbers can have nothing to do with any application of the theory of probability.

We may further, and in conclusion, remark on yet another use of the term *random* in experimental situations. An experimenter employing biological material may find himself in a position where he wishes to be able to assure his colleagues that he has not gone out of his way to employ aberrant material or circumstances, specially suited to the establishment of a preconceived thesis. He may summarise the reasonable precautions he has taken with this end in view by saying that he is employing material selected at random, or that it comprises a random sample from some ill-specified population. From the present point of view all that he can legitimately mean by this is that his procedure in the choice of experimental material has been honest or unbiassed. In the case of static randomness we have been able to point to some sort of formal rationale underlying the procedures involved. In the present case we cannot reasonably hope for a corresponding development, such as would be subsumed in a calculus designed with the function of guaranteeing mere good intent.

7. Costing of Information Potential

We continue in our final chapter the discussion of information potential commenced in Chapter 2. In comparing two coding systems, introduced say for purposes of error-correction, we have preferred the system involving the more economical use of the material resources available, specifically of the tape onto which symbols are fed. We may generalise this procedure by supposing that the resources referred to have assigned costs or *cash-values*. If, in particular, we attach cash-values to the symbols an operator can employ, and consider situations in which the operator has a definite cash-allocation available to him in devising an information store, we are led to a formal development very closely parallel to that of classical thermodynamics, but meaningful in its own right. Information potential corresponds to entropy.

There has long been known to be a connection between thermodynamics and information theory, and many writers have drawn a comparison between entropy (or negative entropy) and information, regarding each as having the properties of a substance; indeed at the outset the stochastic source theory drew heavily on the ideas on formalism of statistical mechanics. The present development suggests that there exist important analogies not merely between the mathematical formalism of Information Theory and Thermodynamics, but between the philosophical orientations appropriate to each; and that it is these analogies which particularly call for exploration. We are led to consider, in a wholly speculative vein, the limitations of the viewpoints we have adopted in respect both of probability and information theory, and the directions in which the ideas of must be developed if an information theory is to result which adequately accommodates the limitations on the transmission or storage of information imposed by considerations of physical smallness.

Cash-allocations and Temperature

Let us envisage the situation in which we have one or more information stores, each consisting of a number of cells—which may or may not be localised—and that either a central stock of symbols or a number of separate stocks is available, each symbol having a definite *cash-value*. We regard a particular information store as fully specified when there is associated with it a definite cash-allocation, E, enabling symbols to be *purchased* from the appropriate stock. There will correspond to a particular value of E an amount of information potential which we may denote by $I(E)$. Two possibilities arise here. Either we regard a cash-allocation E as allowing the store concerned to receive messages bearing a cost, in terms of symbols, less than or equal to the value E, or we confine allowed messages to those whose value is exactly E. Of these two alternatives we choose the second, though it involves considerable mathematical difficulties. Before proceeding further we must draw particular attention to these difficulties.

If E is treated as a continuous variate, then in general $I(E)$ will be a very queer function, reflecting in an incomplete and erratic way what one might suppose at root to be a straightforward relationship. We may get round this difficulty by allowing a little latitude in picking messages with nominal value E. If we allow the expenditure of any sum in the range $E + \Delta E$ it will often happen that the amount of information potential which becomes available is very insensitive to variations in ΔE, provided only that it is small and non-zero*. This notion enables us to treat I as a continuous function of E, on the understanding that in a more extensive treatment the matter could be discussed with full mathematical rigour. We may note that the same cannot be said in respect of the thermodynamic analogy which we discuss later. There, intrinsic physical difficulties, relating to the precise assignment of internal energy in an adiabatically enclosed thermodynamic system, inhibit the exact realisation of the mathematical methods employed.

We proceed, accordingly, with a certain amount of caution. Let us think of an operator in charge of a particular information store, which is such that the amount of information potential available to him is $I(E)$, E being his cash-allocation. He may change his stock of information potential by augmenting his cash-allocation. The cost to him *per unit information potential* of increasing his stock by a small amount is

$$T = T(E) = 1 \left/ \frac{dI}{dE} \right. .$$

We refer to T as the *temperature* of the store, and note that if we allow the operator an amount of additional cash, ΔE, there becomes available to him additional information potential $\Delta E/T$. The temperature of a store can either

---

* This phenomenon, which is purely mathematical, is sometimes referred to as the stability of thermodynamic functions. We have already encountered an extreme example (p. 18). If values 0 and ε are attached to symbols in a store comprising n cells of binary tape, then if n is large, the amount of information potential associated with cash-allocation $n\varepsilon/2$ is almost the same as if there were no restriction at all on available cash.

be positive or negative (or zero), though we might think it natural to deal first with situations in which T is positive.

To illustrate the notion of temperature, let us consider the special case of a *binomial store*, that is to say an information store comprising a piece of tape with N (localised) cells into each of which the operator can insert one of two symbols*. Let us suppose that the cost of an individual symbol of the first type is 0, and of the second type is u, and that the operator's cash-allocation is

$$E = ur,$$

so that, in effect, he has at his disposal $N - r$ symbols of one kind and r of another.

The information potential of the store, using natural units, is therefore

$$I = \ln \binom{N}{r}.$$

If both N and r are large, we have:

$$I = - N \left\{ \frac{E}{uN} \ln \frac{E}{uN} + \left(1 - \frac{E}{uN} \right) \ln \left(1 - \frac{E}{uN} \right) \right\}.$$

The temperature of the store is

$$T = 1 \bigg/ \frac{\partial I}{\partial E} = u / \ln \left(\frac{Nu}{E} - 1 \right).$$

As E increases from the neighbourhood of 0 to the neighbourhood of Nu, I increases from zero (one message only possible: a string of zero-value symbols) to a maximum of $N \ln 2$ when $E = Nu/2$, and decreases thereafter symmetrically to zero. This we have already, in effect, noted on p. 18. T increases from zero at $E = 0$ to ∞ at $E = Nu/2$, switches abruptly to $- \infty$, and thereafter increases, to become zero when E achieves its maximum value of Nu.

This progression, that is to say from zero through positive values to $+ \infty$, then from $- \infty$, through negative values to zero again, represents a natural ordering, from what we can think of as *low* to what we can think of as *high*, for cost-entities of which information-temperature is an example, as the following considerations indicate. Let us suppose that there are two operators, each with a large information store and a definite cash-allocation; and that we allow them to co-operate by interchanging cash, but not otherwise. Suppose

---

* For the corresponding set-up in the thermodynamic analogy, see Schrödinger, E. (1946). "Statistical Thermodynamics", p. 20.

E

that operator A, whose store has temperature T_A passes over cash ΔE to operator B, whose store-temperature is T_B. The overall change in information potential is

$$\frac{dI_B}{dE}\,\Delta E - \frac{dI_A}{dE}\,\Delta E = \Delta E\left(\frac{1}{T_B} - \frac{1}{T_A}\right).$$

Since this expression is meaningless when either temperature is zero, we can conveniently regard the extreme low and high temperatures as an arbitrarily small positive number and an arbitrarily small negative number respectively. The expression is positive if and only if T_A is higher up the scale of temperature which we have indicated than T_B; so that combining two information stores having different temperatures results in an increase of total available information potential if cash is transferred from the higher temperature store to the lower temperature store. Alternatively, we can say that the transfer of a small amount of cash from a high temperature store to a low temperature store necessarily results in an increase in overall information potential.

Cash Released by Combining of Stores

In the previous section we have envisaged situations in which information stores originally separate are joined together to provide a single source of information potential, their separate cash-allocations being pooled. An increase in information potential usually occurs, and a decrease certainly cannot occur, since the operator in charge of the combined store is free to disregard the amalgamation and utilise the component stores independently. We may, however, look at the amalgamation procedure from a quite different point of view. Suppose the two operators choose to co-operate, not in order to increase the overall information potential, but in order to release cash, as it were for other purposes, having been granted their original cash-allocations in order to make available an assigned amount of information potential. We can enquire as to what ratio the cash which B can release bears to a small amount of cash transferred to him by A.

If T_A and T_B are both positive, or if T_A is negative and T_B positive, this ratio is

$$1 - \frac{T_B}{T_A},$$

which is unity if T_B is zero, and zero if $T_B = T_A$; if T_A is negative, more cash is released than transferred*. If T_A and T_B are both negative and $|T_A| < |T_B|$, cash must be transferred from A to B to increase information potential, and from B to A to release cash. We may note that, qualitatively, the underlying

* With a physical analogy in mind, we might refer to this as the *laser effect*.

idea does not depend on the stores being infinitely large, nor on temperatures being definable. If a cash allowance is spread over two stores previously with independent allocations. we can, quite generally, look at the matter in two ways: we can either think of the information potential as increasing, the total cash-allocation remaining constant, or we can think of the original information potential being retained and a calculable amount of cash, in consequence, being released.

We can pursue the same line of thought in another direction. Suppose we have a store in which the cost of individual messages depends on some external parameter or parameters, controlling the costs of different symbols; a simple case might involve something analogous to a revaluation of currency. A change in such a parameter results in an alteration in the information potential, if we suppose the cash allowance to remain constant. In the alternative interpretation, in which we require the available information potential to remain constant, a change in an external parameter either releases cash or requires the provision of more cash. In the extreme case, where parameters are adjusted so that the cost of all messages becomes zero, the whole of the allocated cash is released.

Small Sub-stores

Consider now the following special situation. Suppose we have an information store with no cash-allocation specified, but capable of absorbing symbols with given values. Such a store will have a value to an operator already in possession of an information store with a cash allocation. This value can be measured in two alternative ways; as the increased information potential which results from its use by the operator in connection with his original store; or as the cash released if the operator wishes to utilize the new store, maintaining the information potential originally available to him. This we can say quite generally. The situation becomes more manageable if we can suppose that the original store possesses an information potential, I, which can be regarded as a continuous function of the cash-allocation E, and if it is so large relative to the store to be appended (which we can refer to as the *sub-store*) that its temperature, T, is unaffected by the connection of the two stores with another.

In this case, let us suppose that the sub-store can accommodate one of the n messages with values $\varepsilon_1, \varepsilon_2, \ldots, \varepsilon_n$. It will be convenient to think of ε_1 as zero. Since the logarithm of the number of messages admissible into the main store in its original state is $I(E)$, the actual number of messages admissible with cash-allocation E is

$$\exp\left[I(E)\right].$$

When the sub-store is appended, the total number of possible messages, given the same cash-allocation, is

$$\sum \exp \left[I(E - \varepsilon_i) \right].$$

Since we are supposing that the ε_i are small compared with E, this expression is equal to

$$\sum \exp \left[I(E) - \varepsilon_i/T \right] = \exp \left[I(E) \right] \sum \exp \left(- \varepsilon_i/T \right).$$

It follows that the increase in the information potential is

$$\ln \sum \exp \left(- \varepsilon_i/T \right),$$

and the cash which would be released if the information potential were kept constant is

$$T \ln \sum \exp \left(- \varepsilon_i/T \right).$$

Each of these expressions expresses the value of the sub-store to the operator. Each depends on the nature of the store into which the sub-store is being incorporated but, as it happens, on no attribute of it other than its temperature.

We can look at the same matter in a slightly different way by regarding the sub-store as embedded in the main store, and by asking what happens when we make some change in the ε's. We define

$$F' = - T \ln \sum \exp \left(- \varepsilon_i/T \right).$$

If we change the specification of the sub-store (for example by altering the external parameters controlling the cost of sub-messages) in such a way that

$$\Delta F' < 0,$$

then cash is released. F' for a sub-store corresponds to E for a complete store: *cf.* the trivial proposition that a change in external parameters of a store which lowers the cost of messages releases cash. This is the reason for introducing the negative sign in defining F'.

We may further note that the number of messages in the overall store which contain the message with value ε_i in the sub-store is

$$\exp I(E - \varepsilon_i) = \exp \left[I(E) \right] \exp \left(- \varepsilon_i/T \right),$$

where $I(E)$ is the information potential as function of cash-allocation for the total store less sub-store, which we refer to as the main store. It follows that the proportion of possible overall messages containing the ith sub-message in the sub-store is

$$\frac{\exp\left(-\varepsilon_i/T\right)}{\sum \exp\left(-\varepsilon_i/T\right)} .$$

This implies what one would expect: expensive words are the least useful. We may obtain this same result in a more elementary and rigorous fashion by using as our main store a binomial store (see Appendix to this Chapter).

Reverting to the main argument we note that the situation becomes even simpler if the sub-store, though still small relative to the main store, is large enough to possess a well-defined information potential which is a function of an independent cash-allocation. In such a situation we can assume that there will be little difference between the total store and the compound store obtained by separating the sub-store from the main store, allocating to the former the amount of cash which gives it temperature T, assigning the remaining cash to the main store, and using the two stores simultaneously; that is to say, we can assume that it makes little difference, as far as information potential is concerned, if we raise or lower a barrier between the main store and the sub-store. The sub-store can now be taken to have a definite information potential, I', as well as a definite cash-allocation, E'. We can suppose the main store to have information potential $(I - I')$ and cash-allocation $(E - E')$.

When attached to the main store the sub-store will, in effect, attract a share of the cash-allocation which gives it temperature $T = T'$, i.e. E' will be such that

$$\frac{dI'}{dE'} = \frac{1}{T}$$

An amount of information potential, I', will now no longer have to be provided by the main store. The cost of this would have been TI', so that the total cash released by the attachement of the sub-store is

$$TI' - E' = T'I' - E'.$$

An extension of these procedures identifies E' with the average cost of sub-messages:

$$E' = \sum \varepsilon_i \exp\left(-\varepsilon_i/T\right)/\sum \exp\left(-\varepsilon_i/T\right);$$

using the relationships

$$I' = \frac{E' - F'}{T'}, F' = -T \ln \sum \exp\left(-\varepsilon_i/T\right),$$

it follows that

$$I' = - \Sigma p_i \ln p_i,$$

where

$$p_i = \exp (- \varepsilon_i / T) / \Sigma \exp (- \varepsilon_i / T)$$

is the proportion of allowable messages in the total store which contain the sub-message with value ε_i. The mode of derivation of this formal expression is of particular interest, since in the Information Theory of Shannon and his disciples it occupies, with the appropriate interpretation, the position of a postulate.

Correspondence with Thermodynamics

As indicated earlier, these is recognisable a close correspondance between the formalism developed in the last three sections and the standard formalism of equilibrium statistical thermodynamics. If, in particular, we confine our attention to those portions of the argument which do not contain explicit reference to individual symbols and their individual values, we see a close correspondence with classical thermodynamics, *i.e.* equilibrium thermodynamics developed without regard for happenings at the molecular level. We can make this correspondence fully explicit by laying out in parallel columns the information-theoretical concepts we have introduced and their thermodynamic analogues, the left-hand column relating to information theory, the right-hand column to thermodynamics.

Information store	Adiabatically-enclosed thermodynamic system
Information potential	Entropy of adiabatically-enclosed system
Cash	Energy
Cash-allocation	Internal energy
Cells	Molecules, electrons etc. (microsystems)
Symbols	Energy-levels
Parameters governing values of symbols	State-variables controlling energy-levels (volume, magnetic field strength, etc.)
Change in required cash-allocations resulting from change in symbol-values	Work done in adiabatic compression or expansion, adiabatic magnetisation or demagnetisation, etc.

Proportionate release of cash by cash-transfers between stores	Efficiency of ideal heat engines
Main store in compound store	Heat-bath
F'	Helmholtz free energy
I'	Entropy in non-isolated system

This catalogue is neither systematic nor complete. We add the following brief notes by way of explanation and amplification.

(i) There is omitted in this list the key correspondence implicit in our use of the thermodynamic term *temperature* for marginal cost of information potential. We may note that negative temperatures, which emerge quite naturally in the simplest discussion of costed information stores, escaped attention in the literature of thermodynamics until comparatively recently*.

(ii) The so-called *Laws of Thermodynamics* appear in the information context either as definitions or tautologies. Thus the introduction of the Cash metaphor subsumes an analogue of the Conservation of Energy. We have made brief reference to the analogue of the Second Law on p. 78, line 20.

(iii) In discussing the qualities of small information stores, that is to say stores in which the amount of cash employed in respect of individual messages is subject to substantial fluctuations, we have throughout regarded such stores as embedded in another store so that the total store has a fixed cash-allocation. In the thermodynamic context this amounts to asserting the primacy of the *microcanonical ensemble*, from whose properties those of the *canonical* ensemble are to be formally derived.

(iv) In considering small information stores relative to entities like temperature definable only in *large* stores, two courses are open. What we chose to do was to attach the small store to a large one, examining what effect this had, for example in increasing information potential. Equally, we could have considered the properties of the store obtained by indefinite replication of the small store. These alternatives admit of parallels in alternative developments of statistical mechanics.

Methodological Implications

The course we have been led to follow in discussing costed information stores corresponds in broad outline with that adopted, and in part initiated, by Willard Gibbs in respect of statistical mechanics. Our systematic use of information potential as the fundamental concept of information theory,

* See Ramsay, N. F. (1956). Thermodynamics and Statistical Thermodynamics at Negative Absolute Temperatures, *Phys. Rev.*, **103**, 20, Bazarov, (1964). "Thermodynamics", Pergamon Press.

however, suggests an interpretation of entropy which differs from that which
an interpretation of entropy which differs from that which Gibbs and many
later writers have favoured. For Gibbs, entropy was essentially a measure of
disorder.* A stochastic interpretation of entropy corresponds to a stochastic
interpretation of *information*; and this latter notion we have rejected for good
reason. We obtain an analogy consistent with our earlier discussion by regarding entropy as a measure of *accessibility*†, and the strictly formalisable part of
traditional thermodynamical theory as being concerned with the apparatus
employed in heat experiments, rather than with what goes on within the
apparatus. This attitude accords with the philosophical view expressed at the
beginning of Chapter 1 of the present book, and arguably with the actual
practice of working scientists in using thermodynamical theory. It places a
severe limitation on extensions of the theory to non-equilibrium situations.

The historical development of thermodynamics has been dominated by the
view that logical precedence must be given to entities, such as temperature
(as contrasted, say with entropy) which seem to be directly measurable, and to
concepts such as irreversibility (as contrasted with, say, isentropy) which
seem to have an immediate physical connotation in terms of the behaviour of
matter. That is to say the underlying philosophy has been essentially Cartesian
in assuming that the purpose of the theory is to provide a direct representation
of what goes on in a real external world. Aside from more fundamental
objections, this is inconsistent with the fact that the correspondence established between fact and theory is arrived at either by the use of elaborate
experimental procedures or by confining attention to highly selected phenomena which are only interpretable by reference to such procedures.

We may introduce here a fiction which corresponds in our scheme of
analogy with the historical derivation of thermodynamics. We have developed
in the preceding pages a calculus which relates to the manipulation of information stores according to defined criteria of economy, and we have examined
in some detail the relationships which exist between the quantities of information potential available in these stores; also their dependence on formal
resources available and on formal manipulation of the hypothetical structures
involved. Let us imagine that in a real world in which physical apparatus is
actually employed for purposes of information storage, a position has been
reached where people actually perform the manipulations we have envisaged
in our abstract discussion. That is to say they utilise information stores in
isolation and in combination, in accordance with the fluctuating availability
of certain material resources, and for purposes which we can leave undefined.
Let us suppose, moreover, that we reserve for inspection just those situations

---

* At least according to a note discovered in his papers after his death.

† *Cf.* Guggenheim, E. A. (1949). Statistical Basis of Thermodynamics, *Research*, **2**, 450.

in which attempt is made to conserve these resources with the utmost effort. Alternatively, let us imagine that, as a result of the exercise of much insight and acuity, we are able to perceive situations in Nature (perhaps, for example, involving molecular biological systems) in which something which corresponds to this sort of activity occurs. In either case what shall we actually see? We shall see, for example, messages passing from one store to another when two stores with quite different qualities are amalgamated or amalgamate; and material resources being called upon when restrictions are relaxed. The characteristic *phenomena* will concern what we shall reasonably think of as the flow of information. We might be led to develop a theory whose object would be to mirror as precisely as possible the material events we observe to take place. The theory, we may suppose, would have much in common with our abstract calculus of information stores, though we would derive it by what would appear to be purely empirical means. Thus concepts corresponding to what we have referred to as Information Potential and Temperature would be thought of and measured in terms of material behaviour in naturally occurring situations, or in the idealisations produced by experimental manipulations.

The superficial virtue of such a theory would lie in its empirical origin, and its apparent independence of metaphysical discussion. Therein, however, would also lie its most severe drawbacks, which we may list as follows.

(a) The theoretical interpretation of the flow of information would have to depend on some sort of calculus of uncertainty, that is to say on a theory of probability; and this, being empirical in nature, would be subject to all the difficulties which we have discussed in another context. Notably, its accuracy would be dependent on and commensurate with the limitations imposed by its degree of verifiability.

(b) The basic concepts of the theory would be inextricably associated with an arbitrary type of physical set-up whose initial selection was dependent on historical accident; so that later applications would be unavoidably dependent upon analogical reasoning.

(c) Finally, and paradoxically, the theory would, in the last resort, fail as an instrument for empirical enquiry, since it would be unable to provide a form of language usable in situations too complicated or too ill-controlled for the development of a mathematical model.

These, we may note, are just the faults characteristic of thermodynamics viewed from the entropy-disorder standpoint.

Micro-information

Our separate discussions of probability theory and information theory have throughout been conditioned by two related qualifications. In discussing probability theory we have observed (p. 5) that the classical theory is limited by the requirement that the labelling of symmetries should not interfere with the symmetries themselves, and in discussing information theory we have held in mind the special type of system in which messages comprise symbols placed into cells; that is to say, in each case we have confined attention to set-ups in which we can make a distinction between the label and the labelled. In each case we should anticipate that the relevance of the theory will break down when we reach a physical level where the entities dealt with are, in some sense which at the moment we cannot define, so small that this sort of distinction cannot be preserved. Thus in considering procedures of information transmission, we should recognise two extreme classes: in which messages can be read off repeatedly in successive acts of reception, and in which messages are annihilated by the act of reception; or, in the matter of information storage, of cases in which a message is available for perpetual inspection, and cases in which it can be withdrawn from the store but once. We may refer to the systems involved as macro-systems and micro-systems respectively. In this book we have dealt only with macro-systems.

Attempts have been made in recent years to obtain what would amount to a theory of micro-information by associating theories conceived in terms of macro-systems, with conventional quantum mechanics. There are reasons, however, for supposing that a much deeper problem is involved than this type of approach would suggest.

In the first place, we may note that quantum mechanics as currently expounded already contains the germs of notions which relate to communication procedures. Thus:

(i) The line of thought which traces the origin of the quantum-mechanical uncertainty principle to the influence of the means of observation on what is observed has clear connections with our actual definition here of the micro-information problem.

(ii) Much of the mathematical apparatus of quantum mechanics was originally borrowed from communication technology; notably the operational calculus, systems of orthogonal functions, *and so on*.

(iii) The later emphasis in the mathematical development of the subject on notions of symmetry and distinguishability again suggests an underlying preoccupation with information ideas.

(iv) The characteristic wave-particle duality of quantum mechanics suggests an analogy, or some closer connection, with the separate functions of sender and receiver in a communication procedure; the sending of a message being a wave-type activity, and reception a matter concerned with observing the particle-events which, in the last analysis, are all that quantum theory recognises as coming to the immediate attention of the experimenter as observer.

Such considerations suggest that we should look forward to a complete fusion of information theory with quantum theory, rather than to any less intimate association.

In the second place, there exist epistemological objections to quantum mechanics affecting its most elementary features. These can, perhaps, be glossed over in presenting the theory as merely descriptive of experimentally-observed phenomena, a theory of spectra. They are more obtrusive when the theory is looked at in the present type of context.

(i) All expositions of quantum mechanics, however abstract, are forced to recapitulate in some way the historical development, in which a start is made by thinking in terms of *particles* whose most characteristic properties are subsequently withdrawn. We might hope for an alternative development to result from attributing at the outset a lower limit to the resources required to accommodate finite quantities of information potential, avoiding reference to considerations of material atomism.

(ii) It is widely recognised to be unsatisfactory that in current expositions relativistic considerations have to be introduced after concepts treated as more fundamental have been established. We may suppose that the limitations on resources which must receive explicit mention in a theory of micro-information will involve just those geometrical issues with which the special theory of relativity itself deals.

The second of these points is of particular interest, insofar as relativity theory affords a striking example, further to that of quantum mechanics itself, of the way in which the theoretical physics of the last hundred years betrays an underlying preoccupation with communication ideas and procedures. Thus the prime novel postulate of the special theory, namely that requiring the existence of an upper limit to signalling velocities, specifically concerns acts of communication; and the opportunity which the theory affords for eliminating the ether as a medium onto which messages can be impressed has as intimate a relationship to our definition of the micro-information problem as that of the uncertainty principle itself.

In conclusion we may make the following comment. The purpose of the present book has been to discuss the logical issues underlying the use of notions of probability and information in biological enquiry. The major part of discussion of information concepts is in fact currently taking place within a biological context, actuated by the increasing recognition of the rôle which communication plays at all levels in the organisation of biological systems. We have ended up, however, with a tentative discussion of issues which are purely physical. Much of the biological discussion to which we have referred, however, tends to assume that physics and engineering are in a position to provide the appropriate intellectual tools in respect of much simpler man-made systems. If we are right in supposing that this is not the case, and that a much closer physical analysis is called for, it would not be the first time that Biology has played a part in providing the stimulus for the development of fresh physical ideas.

Appendix (*see p. 81*)

Consider an information store comprising N cells, each admitting one of two symbols with values 0 and 1. Let us append to this store a single cell which can be occupied by one of n symbols with values $\varepsilon_1, \varepsilon_2, \ldots, \varepsilon_n$, each an integer, $\varepsilon_1 = 0$. If the total cash-allocation is E (integral), then the number of disting-uishable messages having ε_i in the $(N + 1)$th cell is

$$\binom{N}{E - \varepsilon_i}$$

The ratio of the number of messages with the ε_i-symbol in the final position to the number with the ε_1-symbol in the final position is

$$\binom{N}{E - \varepsilon_i} \bigg/ \binom{N}{E} = \frac{E!(N - E)!}{(E - \varepsilon_i)!\,(N - E + \varepsilon_i)!}.$$

Let us now set $E = \lambda N$, where $0 < \lambda < 1$, and write $1 - \lambda = \mu$. The above ratio can then be written

$$\frac{\lambda N(\lambda N - 1) \ldots (\lambda N - \varepsilon_i + 1)}{(\mu N + \varepsilon_i)\,(\mu N + \varepsilon_i - 1) \ldots (\mu N + 1)}.$$

If now we let N tend to infinity, keeping ε_i at its constant value, the ratio becomes

$$\left(\frac{\lambda}{\mu}\right)^{\varepsilon_i}.$$

We can write this as

$$\exp\left(-\varepsilon_i \ln \frac{\mu}{\lambda}\right) = \exp\left[-\varepsilon_i \ln\left(\frac{N}{E} - 1\right)\right] = \exp\left(-\frac{\varepsilon_i}{T}\right),$$

where T is the temperature of the binomial store as derived earlier, with $u = 1$.

Index